FEAST
TOP RESTAURANT DESIGN

宴——顶级餐饮空间设计

精品文化工作室　编

大连理工大学出版社

Dalian University of Technology Press

图书在版编目(CIP)数据

宴 : 顶级餐饮空间设计 : 汉英对照 / 精品文化工
作室编. — 大连 : 大连理工大学出版社, 2011.6
　　ISBN 978-7-5611-6268-2

　　Ⅰ.①宴… Ⅱ.①精… Ⅲ.①餐馆—室内装饰设计—
汉、英 Ⅳ.①TU247.3

　　中国版本图书馆CIP数据核字（2011）第106729号

出版发行：大连理工大学出版社
　　　　　　（地址：大连市软件园路80号　邮编：116023）
印　　刷：利丰雅高印刷（深圳）有限公司
幅面尺寸：246mm×280mm
印　　张：23
插　　页：4
出版时间：2011年6月第1版
印刷时间：2011年6月第1次印刷
责任编辑：裘美倩
责任校对：李　楠
封面设计：李红靖

ISBN 978-7-5611-6268-2
定　　价：328.00元

电　话：0411-84708842
传　真：0411-84701466
邮　购：0411-84703636
E-mail：designbooks_dutp@yahoo.cn
URL：http://www.dutp.cn

如有质量问题请联系出版中心：（0411）84709043　84709246

CONTENTS
目录

序 PREFACE

多元化融合

"一个人的成就永远超不过他的眼界"，如果用这句话来描述设计思路可能不会被理解，可把这句话归结到企业成功人士的思维模式就会变得耐人寻味。奢华类餐厅消费的主流是企业家、集团高管、高级白领、海归派、艺术家、社会名流、政府官员等，这就要求设计师对奢华类餐饮空间设计的"眼界"换个新的角度，奢华类餐厅不单单是味蕾艺术的传达、视觉饕餮盛宴的展现，"界而不界"——"心觉"才是餐饮空间更加注重的"眼界"。

风格上无论低调奢华还是雍容华贵，只是我们设计表现上的一种手法。真正寻求设计与商业如何结合的可行性，就应该更多地理解高端餐饮消费模式的架构，由此与餐饮商业模式紧密结合，形成高端餐饮的功能架构、流线架构。餐厅应该加以广义地理解，它是客人生活方式的延续，商务活动的延展。充分考虑到消费过程的链接性，就餐只是过程的一部分（餐前、餐中、餐后有着不同的需求点，要满足客人每一个阶段的消费模式和心理诉求），由此延展出高端餐厅不同功能的需求，也就延展出了跨界概念的设计。

《天津八号御膳》《保定珍逸食神》《扬州天鸿和》三个项目分别换了不同的视角做了"跨界"设计尝试，这是我们基于高端客户生活方式的理解给出的定位。奢华餐厅设计不单是表面奢华装饰而已，奢华的价值源自心理价位，它是出品、环境、服务的综合，透过设计的链接形成高端客户的消费场所，从而形成品牌效应使其与其他高端消费产生共鸣，最终形成20%高端客户所需的品牌商务圈以及对等的餐饮文化圈。

无论是高品质的家宴还是奢华的商务宴请，都是消费模式的一种体现，也折射出高端客户群生活方式观念的"界"定。给奢华类餐厅消费主流的"界"就源自本文开篇之句。进入到客户生活观念的"眼界"为客户打造他们的"圈文化"，找到客户的买点，为高端客户"定制"专属的餐饮生活空间，奢华餐厅设计理念与元素也就顺理成章了，我们会从客户在不同业态消费的模式中寻求跨界的融合，于是餐厅空间设计也就有了多元化的记忆片段，也就多了份熟悉与亲切。融入会所、酒店甚至高尔夫、大型 shopping mall 的商务休闲的经营概念。使客人进入他熟知的"圈文化"，生活圈、朋友圈、休闲圈、商务圈的融合，让客户在餐饮空间中"收藏别样的快乐生活片段"，　环球旅游——航海记忆、庄园情节——怀旧留声机、激情蓝调——爵士乐、法式红酒——异国情节、运动休闲——绿野高尔夫、收藏心情——水晶之恋……跳出餐饮空间范畴的"界定"，在隶属高端客户共性的生活模式中寻求新的设计切入点。《天津八号御膳》融合一线品牌的理念；《保定珍逸食神》寻求品质生活的快乐记忆、《扬州天鸿和》融汇酒店会所的气质；三个典型的空间"跨界"设计深受高端客户的青睐，也使得我们在餐饮空间设计当中去探求更多的可能性。

"空间无界"，设计亦是如此。

北京大石代设计咨询有限公司　吴晓温

设计过程中所要考虑的最重要方面是空间的使用者，也就是项目的区域性顾客。他们在餐厅消费的心态、习惯是什么，都喜好些什么等等都是我们在设计过程中要提炼的东西，在这样的基础上我们能否设计出更好更完美的。这些反映到具体的项目中也就是指空间的规划、材质的使用、灯光亮度及形式等设计重心。　其实餐厅的设计用简单的话概括，无非就是对顾客在现阶段用餐模式上给予更优越的用餐方式及用餐气氛，主要包括灯光、空间形态、音乐、服务流程设计等方面。对于有效处理以上问题，我觉得这样几点思维方式对项目设计者在设计时非常重要：

A. 人对物体的感知总是趋成完美正规工整的思维方式。B. 人对物体的感知总是对物体简化、规整的思维方式，结合自己本身的记忆与经验来间接产生对所看到物体的感觉，我们称之为痕迹事件。行动的痕迹给人对事件的复古（还原）行动的记忆的痕迹，从而产生对事件或物体的感知。例如，沙漠里的脚印总是让人想到人的路过。C. 空间设计成功点是让使用者在不同的时间不同的心情下，有不同的思维（使用者在我们的空间里编故事的想法）方法或感知，就是说我们的设计要有悬念。

徐庄设计 庄仙程

Diversified Fusion

It may not be understandable to describe the thoughts of design using this sentence, "A person's achievement will never surpass his field of vision." However, when it comes to a successful entrepreneur's mind-set, the above words just afford food for thought. The mainstream consumers of luxurious restaurants are entrepreneurs, group's senior management, high-level white collars, overseas returnees, artists, celebrities and governmental officers, etc., therefore, a designer who will design a deluxe dining space has to view his design through a new angle, meaning such restaurants are more than to express the art of taste buds, display visually extravagant gourmets. Boundary yet boundless—"mental consciousness" is nothing but the "vision" a dining space values.

A style, no matter it is low luxury or distinguished elegance, is just a way of expressing the design. To truly seek a feasible bonding of the design and business, we should get more understanding of the structure of the consumption mode of high-end catering to closely ingrate it with the business mode, forming the functional structure, streamline structure of high-end catering industry. The restaurant, which should be given a broader definition, is the continuance of the customer's lifestyle and the extension of his business activities. Taking the links of consumption process into full account, we would find dining is just a part of the process as customers have different requirements for individual period of pre-dining, in-dining and post-dining while we need to satisfy customers' mode of consumption and psychological demands at every stage, then, the need for a high-end restaurant with multi-functions is derived and developed as well as the concept of crossover design.

Tianjin No.8 Imperial Restaurant, Baoding ZhenYi, Yangzhou Tian Honghe are three projects attempting to realize a crossover design through different view angles, which also demonstrate the positioning based on our understanding of high-end customers' lifestyle. A luxurious restaurant's design is not limited to superficial sumptuous decorations as the value of luxury origins from the speculative price level, a combination of product, environment and service. The high-end customers' consumption place takes shape through the link of design, resulting in a brand effect to echo with other high-end consumptions and finally forming a branded business circle and corresponding dining culture circle as required by 20% high-end customers.

Either a top quality family dinner or a luxurious business banquet is an expression of consumption mode, reflecting how high-end consumer groups define their lifestyles. Defining the mainstream consumers of luxurious restaurants responds to the opening sentence of this article, which is done like this: enter the vision of customers' living concept to create their own "circle culture"; discover the commodity customers believe worthy buying and customize a dining life space exclusive to high-end customers. So the philosophy and element for designing a luxurious restaurant become logical. We'll seek a crossover blending from different modes of consumption the customers have in various retail formats, thus, the design of dining space is given diversified memory fragments, making people feel it familiar and cordial. Integrated with business casual elements like club, restaurant, even golf course, large shopping mall into an operation concept, the projects allow customers to enter a "circle culture" he is comfortable with; the marriage of life circle, friend circle, leisure circle and business circle allows customers to collect fragments of extraordinary happy life in the dinning spaces. Global tour—memory of sea voyage, chateau plot—vintage gramophone, passionate blues—Jazz music, French red wine—exotic feel, sports leisure—green field golf, collect one's mood—crystal love….break out of the box defining a dining space, instead, pursue a new cutting point of design from the life patterns common to high-end customers. Tianjin No. 8 Imperial Restaurant combines the idea of a top brand; Baoding ZhenYi seeks joyful memory of quality life and Yangzhou Tian Honghe reflects a hotel club ambience; three typical crossover designs are deeply popular among high-end customers, making us explore more possibilities for the design of dining spaces.

"Space is unlimited", so is the design.

BEIJING DASHIDAI DESIGN&CONSULTING COMPANY Wu Xiaowen

The most important aspect to be considered during the design is the users of the space, namely the regional customers of this project. What their consumption minds are, what their habits are, and what they are fond of, etc, are what we need to refine during the design. What we also need to consider is whether we can design a better and more perfect one on such a basis. These can be reflected on the detailed projects, namely the space planning, the use of material, lamp brightness and forms, etc.

In fact, the design of dining room, in a nutshell, is to give customers more superior dining manner and dining atmosphere based on current dining manner, mainly including lamplight, the spatial shape, music, service process design, etc. To effectively deal with above problems, I think the following ways of thinking for the project designers are of great importance:

A. People's perception of the object always tends to be a perfect, formal, and neat way of thinking.

B. People's perception of object is always to simplify the object and to form a neat ways of thinking, combined with their memory and experience to indirectly form a feeling about what they see, which we call Trace Event.

Action trace is about people's memory trace to events' reductive action, and then gives birth to the perception of events or objects.

For example, Footprints on the desert always make people think of someone passing by.

C. The successful point of the spatial design is to let users have different ways of thinking (the ideas that users make up stories in our space) or perception, in different time, under different moods. It means our design should be full of suspense.

Designer for Xuzhuang: Zhuang Xiancheng

奢华欧式

European Luxury

FORWARD in Beijing

北京丰沃德

设计公司：SAKO 建筑设计工社
设 计 师：迫庆一郎
项目地点：中国北京
建筑面积：1050 平方米

FORWARD in Beijing is a Chinese Restaurant with undulated lattice partition. The knitted wood veneer in lattice pattern is undulated horizontally and vertically, creating the partition with both flexible and rigid characters. The partition creates three different spatial layers – opened sofa seating area at the windows, intermediate main dining area and closed private rooms. Fish tanks and liquor cabinets consist of water, glass, mirror and mirror finish stainless steel greet guests at the main entrance.

北京丰沃德是一家中式餐厅，以波状的木格子隔断为特色。这些
网格饰面的格子纵横交错，规则稳定中体现灵活多变。隔断形
成三层空间——靠近窗户的开放沙发座位区、中间的主用餐区以及
较封闭的私人包房。由水、玻璃、镜面及镜面抛光不锈钢组成的金
鱼缸及酒品陈列柜在主入口迎接着宾客的到来。

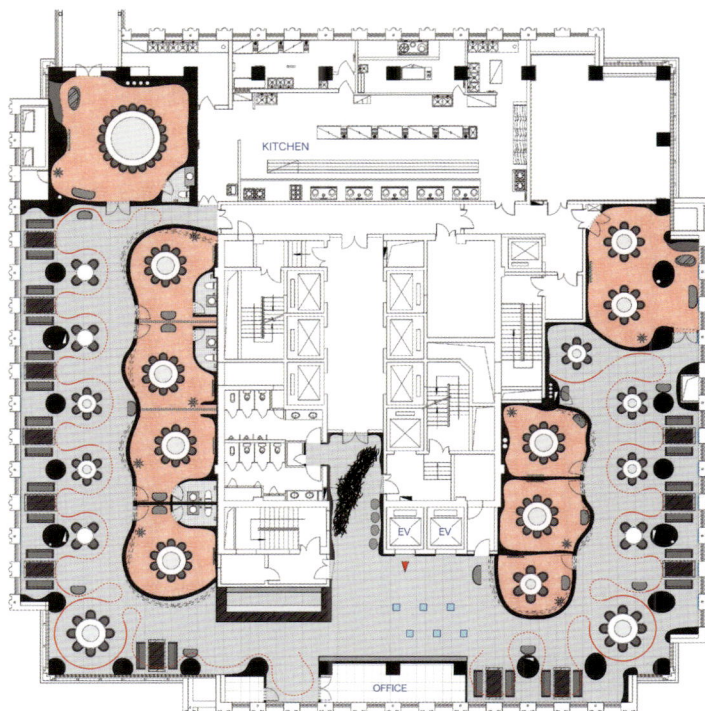

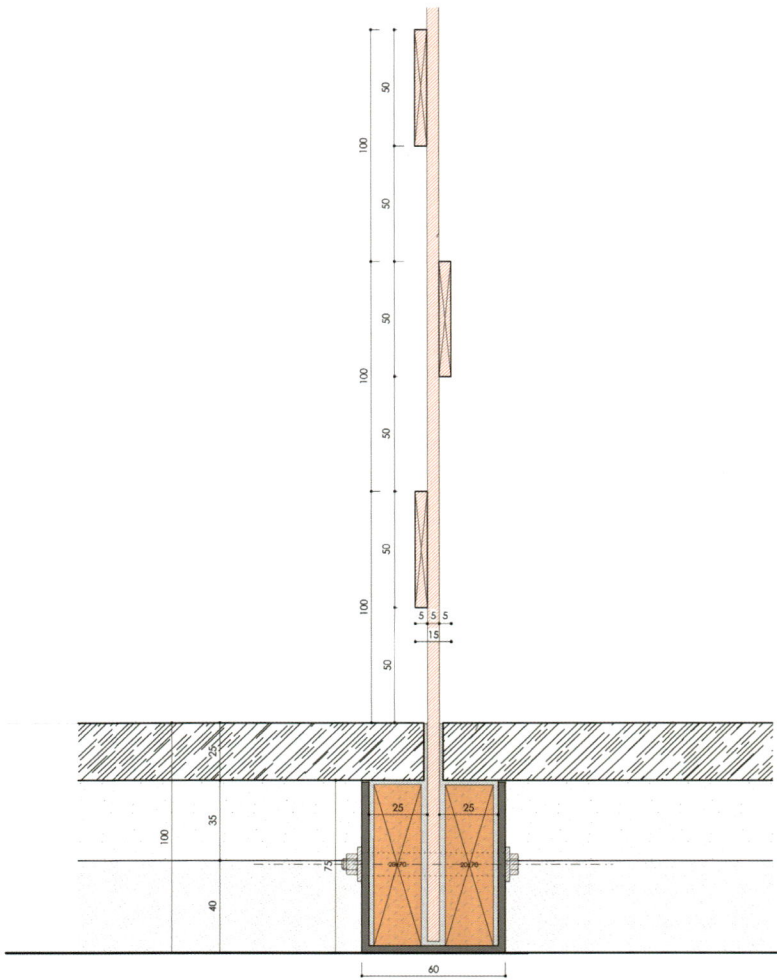

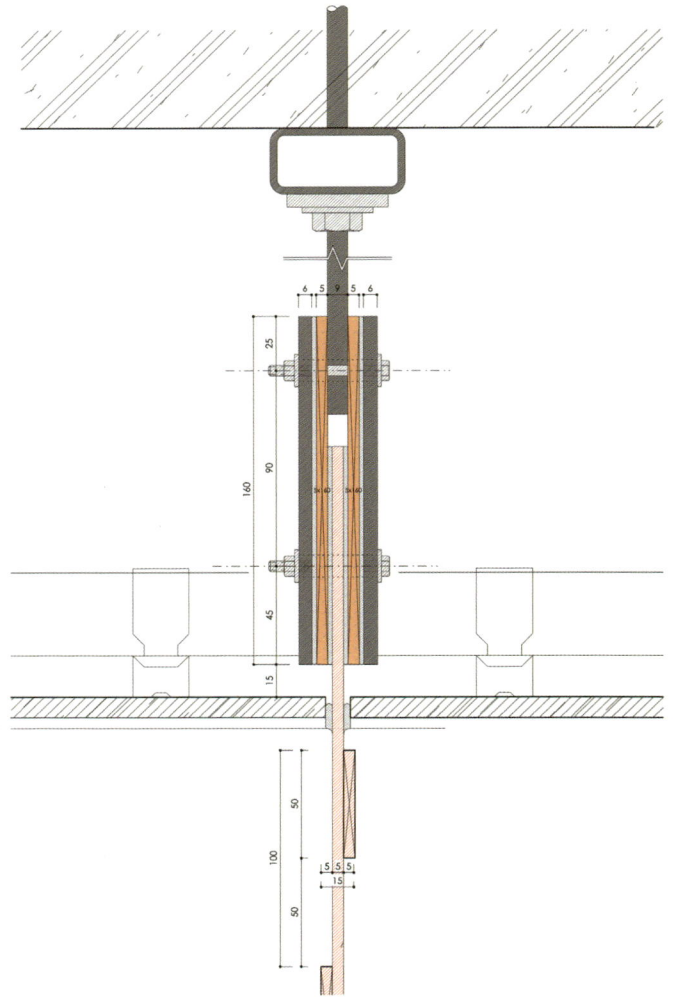

Tianjin No.8 Imperial Restaurant

天津八号御膳

设计公司：北京大石代设计咨询有限公司
设 计 师：吴晓温 张迎军
项目地点：天津市南开区城厢中路
摄　　影：邢振涛

Tianjin No.8 Imperial Restaurant adopts Shanghai noble recreational style in order to adapt to the businessmen's personage psychological demands in the newly developed business area. It emphasize the catering culture, and diversified and aesthetic value of the furnishings, fully embodying the culture and lifestyle of a brand. As a top restaurant, Tianjin No. 8 Restaurant plays more the role of a business platform, which also attracts high-quality banquets.

In the vaulted corridor of the lobby, an exaggerated proportions of chess, a remote blue water reception desk are for decoration. There aren't too much modeling or gorgeous lights, so as to make the customers relaxed and be ready to enjoy the dinner……

The compartments have different themes to cater for different lifestyle, the target customer groups are mainly businessmen, bosses and famous enterprise leaders, government dignitaries. While at the same time, the ordinary customers can be served in the hall to compensate. In the layout, the design considers for the privacy and noble taste(The major customers around are top and medium end customers, so the hall is used to serve them). Catering, business association and leisure consumption are provided at the same time. The restaurant not simply offers catering, but also incoporates business operating ideas of clubs, hotels and even golf. The diversified design ideas create unique business dining and elegant leisure environment.

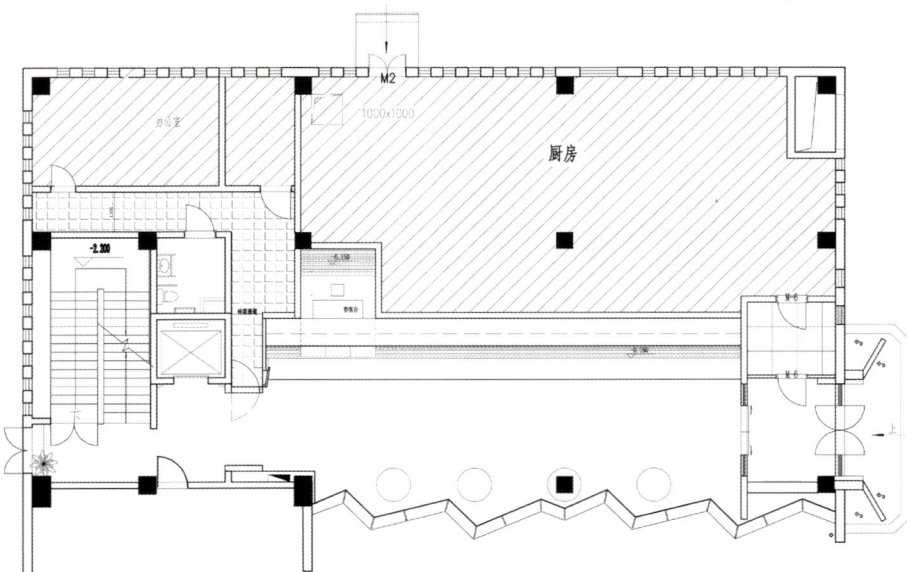

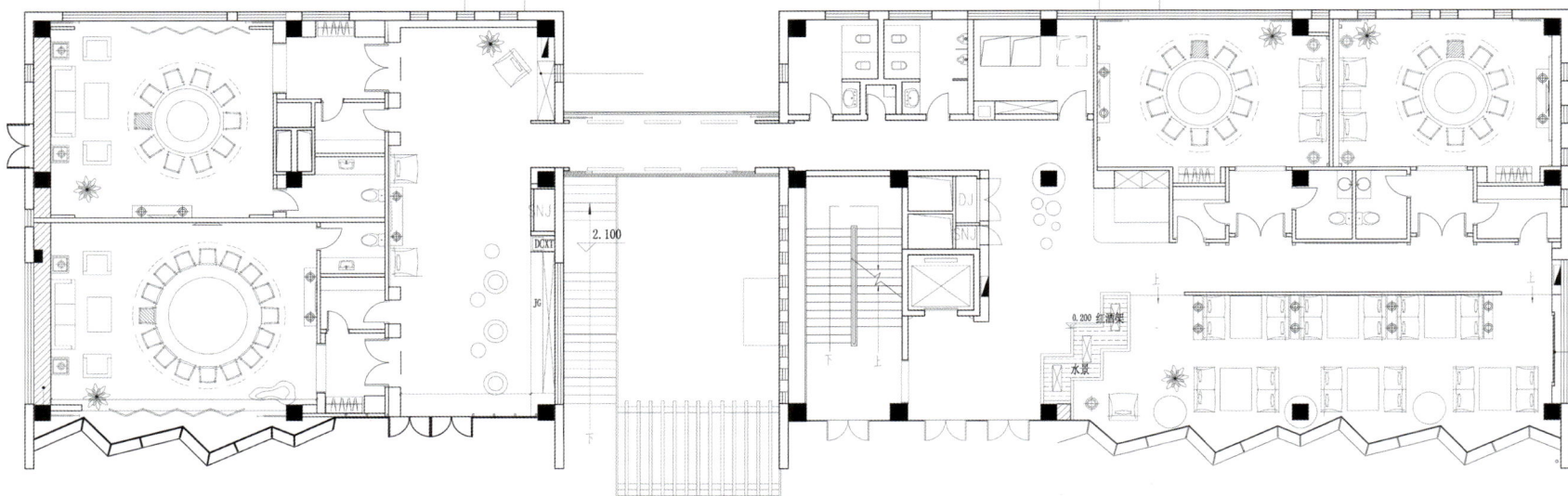

天津八号御膳在设计上采用海派新贵的休闲风格，以适应新兴商业区商务人士的心理诉求，强调餐饮文化内涵，重视家具陈设的多样变化和艺术的观赏性，充分体现来自一线品牌的文化和生活方式，作为高档餐厅的天津八号御膳更多的是起到了商务平台的作用，同时也吸引着高品质的家宴。

大堂拱形长廊中制作夸大比例的国际象棋、幽蓝的水面隐退的接待台，长廊设计没有过多琐碎的造型及绚丽的灯光，目的是让客人以简单的心境使身心放松下来准备接下来盛宴的开始……

别具特色的包间以不同的主题演绎别样的生活方式，消费群体侧重商务人士、公司老板和著名企业负责人、政府机关政要，同时以散厅消费来弥补中端消费客户，注重散厅布局形式的私密性、品味性（周边餐厅多定位中端、中高端，散厅区域可吸纳中高端客户），餐饮与商务交际、休闲消费紧密结合，不单纯为做餐饮而做单一的心理感受，融入会所、酒店甚至高尔夫的商务休闲的经营概念，多元化的思路打造独特的商务用餐休闲雅致的环境。

Jin Xiangyu Restaurant

金湘玉餐厅

设计公司：Nota Design International Pte Ltd
设 计 师：Keat Ong, Gary Zeng, Zhigang Sun
项目地点：沈阳
项目面积：8500 平方米
摄　　影：Jian Long

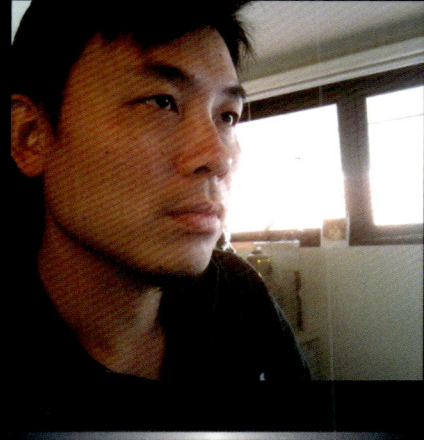

This is a design for a 2-storey high-end contemporary restaurant in Shenyang with a total built-in area of 8500 sqm. It serves western and eastern fusion delicacies.

The restaurant consists of a few important primary areas and programs namely, the Main Dining Hall (with light stage), the Side Dining Hall (with main stage), VIP Room 1, VIP Rom 2, VIP Room 3, Buffet Counters, Serving Counters, Seafood Counter, Entrance Hall (with waterscape features), Service Storage Zone and Kitchen Zone.

The Main focus of the interior architecture would have to be the Front of the Restaurant, frosted by a huge glass curtain wall. The triple volume space and full height glass curtain wall allow us to use inetrior design to contribute to the total architecture of the building. Recessed alcoves with red light shades are placed in a random yet orderly formation with black granite strips of different thickness as the main cladding materials.

The floor is divided into seperate programs, where dining hall is decentralised. Natural scapes are introduced to help demarcate the zones without visusal blockades. More importantly, every public zones have their own special focal points and conversational pieces to ensure not only a gastronomic feast, but a visual feast for the diners!

这是沈阳一家高端的两层现代餐厅，该餐厅总面积 8500 平方米，同时兼顾了东西方两种品味。

餐厅有一些重要的主打领域和项目：主宴会厅、侧宴会厅、VIP 一房、VIP 二房、VIP 三房、自助柜台、服务台、海鲜柜台、门廊、服务储备区和厨房区域。室内装修的主要焦点在餐厅前，有一面巨大的磨砂玻璃墙。三重空间和全高的玻璃墙使室内的总装修十分出彩，发红光的柱子都放在黑色花岗岩带随机而有秩序地形成。餐区随着地板被分为单独的空间。自然景观是为了帮助划定没有视觉障碍的区域，更重要的是，每个公共区域都有自己特别的焦点和谈资以确保不仅是美食的盛宴，而且也是视觉盛宴！

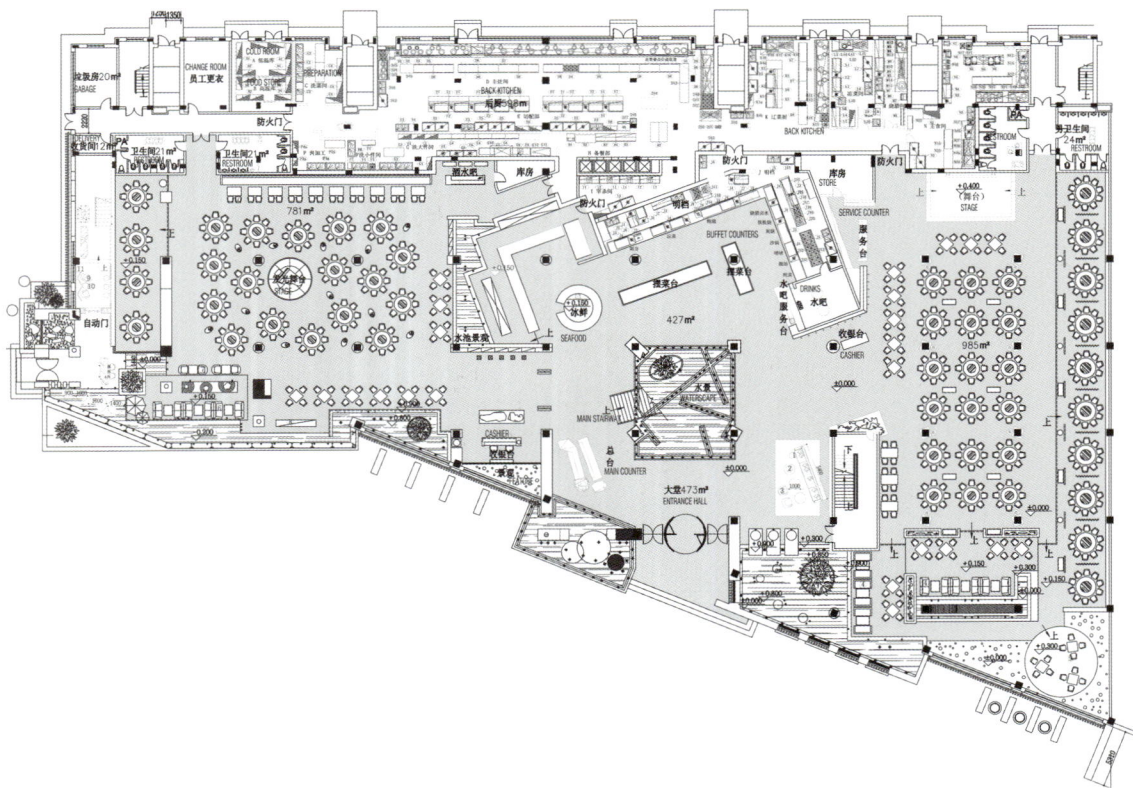

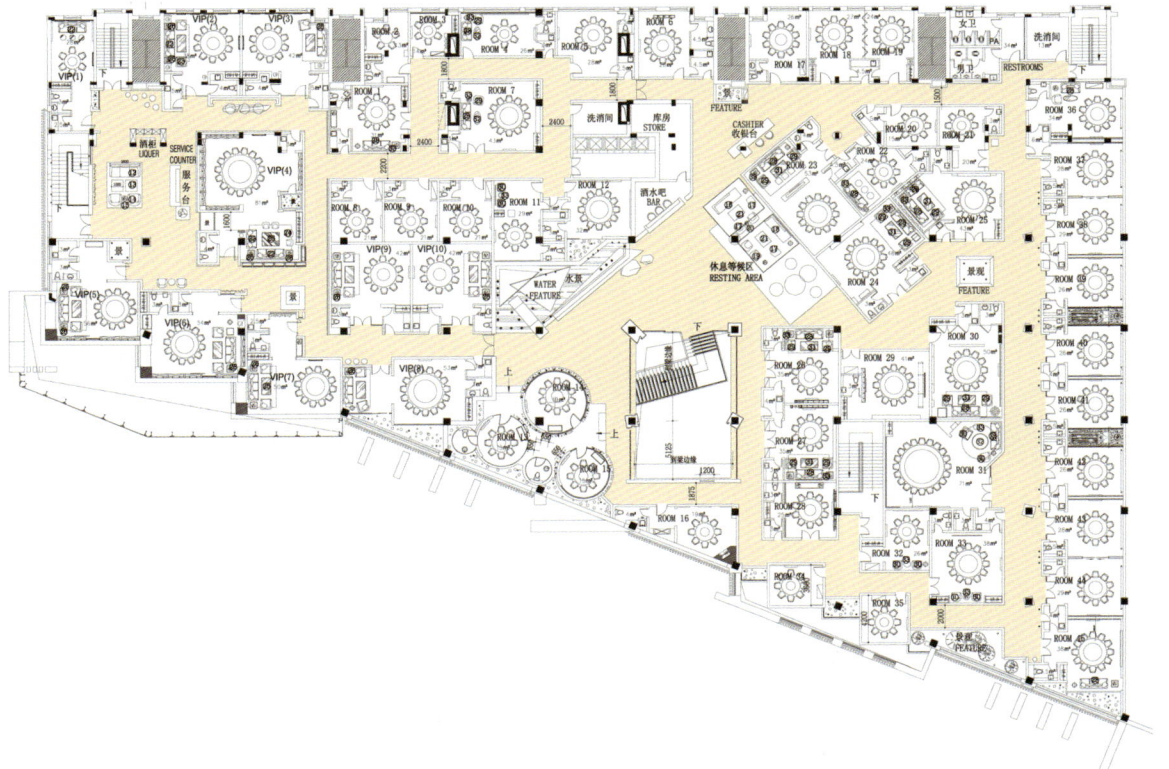

Ruyi Club

如意会所

设计公司：睿智汇设计公司
设 计 师：王俊钦　彭晴
参与设计：赵文静　曹永辉
项目地点：北京市鸟巢西侧盘古大观七星摩根广场
项目面积：500 平方米
主要材料：镜面、拉丝玫瑰金不锈钢、牛皮、银箔、金箔、壁纸、茶镜、
　　　　　明镜、橡木饰面板、西班牙米黄石材、雅士白石材
摄　　影：孙翔宇
文　　稿：朱毅

This case aims to design a club. The center hall is connected with three exclusive VIP rooms, which are independent of each other and they reveal personal temperament. "Ruyi" (which in Chinese means realizing one's wish) is the design theme, "auspicious clouds, ganoderma lucidum, ruyi". There's curling flowers and leaves, which are endowed with the spirit of auspicious cloud.

As for the design of the corridor, aisle design, drawing rose gold stainless steel building modeling signage, exquisite cowhide type curved surface and gold foil type dome, pierce white stone ground, all demonstrate the richness and status of the customers. Entering the hall, the heaven silver top space creates magnificent scenes, which stunns the stadium. The ceiling with LED fiber optic light builds the effect of flowing clouds and running water, the lines of auspicious clouds and running water of "ruyi" outline a low-key luxury, and the cow leather wall surface is presented by the metal buckles with the imagery pattern-carving, which adds more sedateness to the space. The waiting room of the hall, the crystal pendant light, red-wine curtain wall, priceless antiques, the fireplace of the wall, and the top furniture of European style, stepping inside, you seem to pass through the time tunnel and enter European aristocracy salon. The three VIP rooms are private and with distinctive characters. It's a gathering place for famous and rich people. This case focuses on Chinese style. The space is divided into receiving area and dining area. The space is open. The design is not traditional Chinese one, but is presented in a simple and steady way. The ceiling is decorated with the gorgeous crystal droplight, which make the space brilliant and colorful. The Metope is decorated with ebony and ChaJing texture which make the space more attractive.

The Italian style VIP room focuses on the Italian luxury and romance. Functionally it is divided into a cigar visitor area and a dining restaurant. The ceiling is decorated with exceeding luxury line carve patterns and is designed on woodwork and tinsel collocation, crossing the dotted with metope enblock leather. Matching the wall surface with engraved designs of line and board, and gold-mirror modern technique, the complicated design becomes simplified. Furniture's lines are simplified by the most delicate design language, and its elegant modeling and gracious and sedate temperament has become the best annotation of Italian style.

French style VIP rooms are featured with the loyal luxury, and the noble luxury and the popular bold and enthusiastic elements are matched so well that they produce impressive charm.

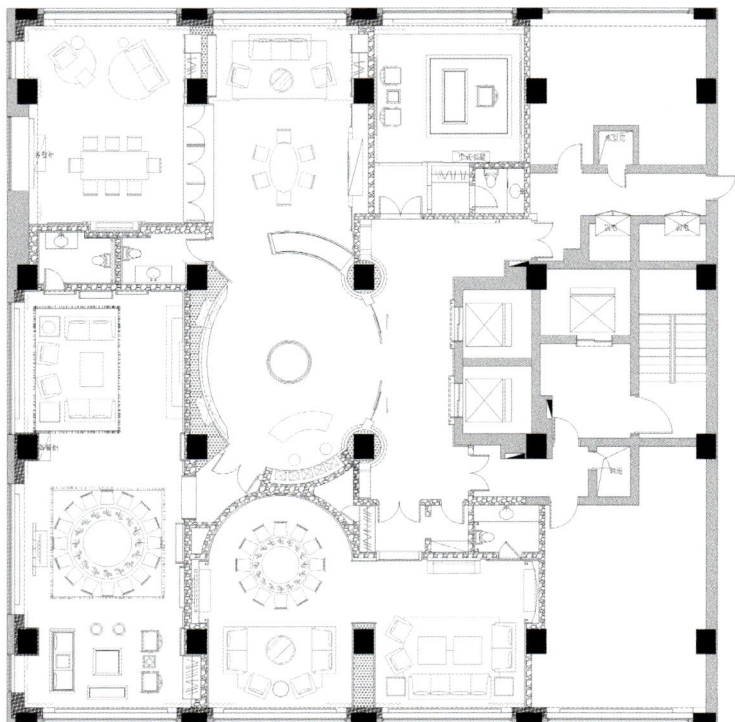

本案以会所形式为设计主轴，以中心大厅连接三大专属贵宾室布局全盘，贵宾室彼此间独立而至，彰显私人气质。以"如意"为此案设计主精神，"祥云、灵芝、如意"，它那旋绕盘曲的似是而非的花叶枝蔓确得祥云之神气。

走廊过道的设计上，拉丝玫瑰金不锈钢的建筑造型门楣、细腻牛皮式曲面、金箔式穹顶、雅士白石材地面，无一不默默地彰显出入此间的客人的富贵。进入大厅，圆融之天穹银顶空间，大气的场面，震慑全场。天穹之处以 LED 光纤灯营造出行云流水之效果，并以"如意"祥云流水之线条勾勒出低调奢华，墙面以金属扣用意象雕花呈现于牛皮墙面，更增添空间之稳重。大厅处的等候区，水晶吊灯、红酒幕墙、价值连城的古董、墙面壁炉、欧式顶级家私，步入其间，仿佛穿越时光的隧道，进入了欧洲的贵族沙龙。

　　大贵宾室个性鲜明，私密性极好，是名流贵宾的社交聚会之处。中式贵宾室为本案中心，此空间以中式设计为主，空间分为会客区及用餐区。空间为开放形式呈现，整体设计并非以传统中式表现，而以简约并稳重的方式展现。室内吊顶用银箔面叠加并旋转，配合华丽的水晶吊灯，把空间装饰得流光溢彩。墙面以黑檀木与茶镜虚实表现稳重质感并起到画龙点睛的作用。

　　意式贵宾室的设计以意式奢华和浪漫之设计思考为主轴，功能上分为雪茄会客区及用餐厅。吊顶以极度奢华的线条雕花及金银箔搭配，交叉点缀着墙面的整体式皮革。搭配墙面、线条线板刻花、金镜现代手法，将繁复之设计简洁化。家具以最精致的设计语言简化线条，其典雅的造型和雍容大度的气质，成为了意大利风格最好的注解。

　　法式贵宾室以巴洛克宫廷奢华风格为主，贵族式的豪华奢丽与流行中的大胆、热情因素的完美混合让法式贵宾室产生令人震撼的魅力。

Nanshan Restaurant in Yuyao

余姚南山饭店

设计公司：宁波市高得装饰设计有限公司
设 计 师：范江
参与设计：卢忆、崔峰
项目面积：2000 平方米
主要材料：镀铜不锈钢、宝丽板、镜面铝塑板、微晶砖、中国黑花岗石、金属镂花板、贴膜玻璃
撰　　文：洪堃

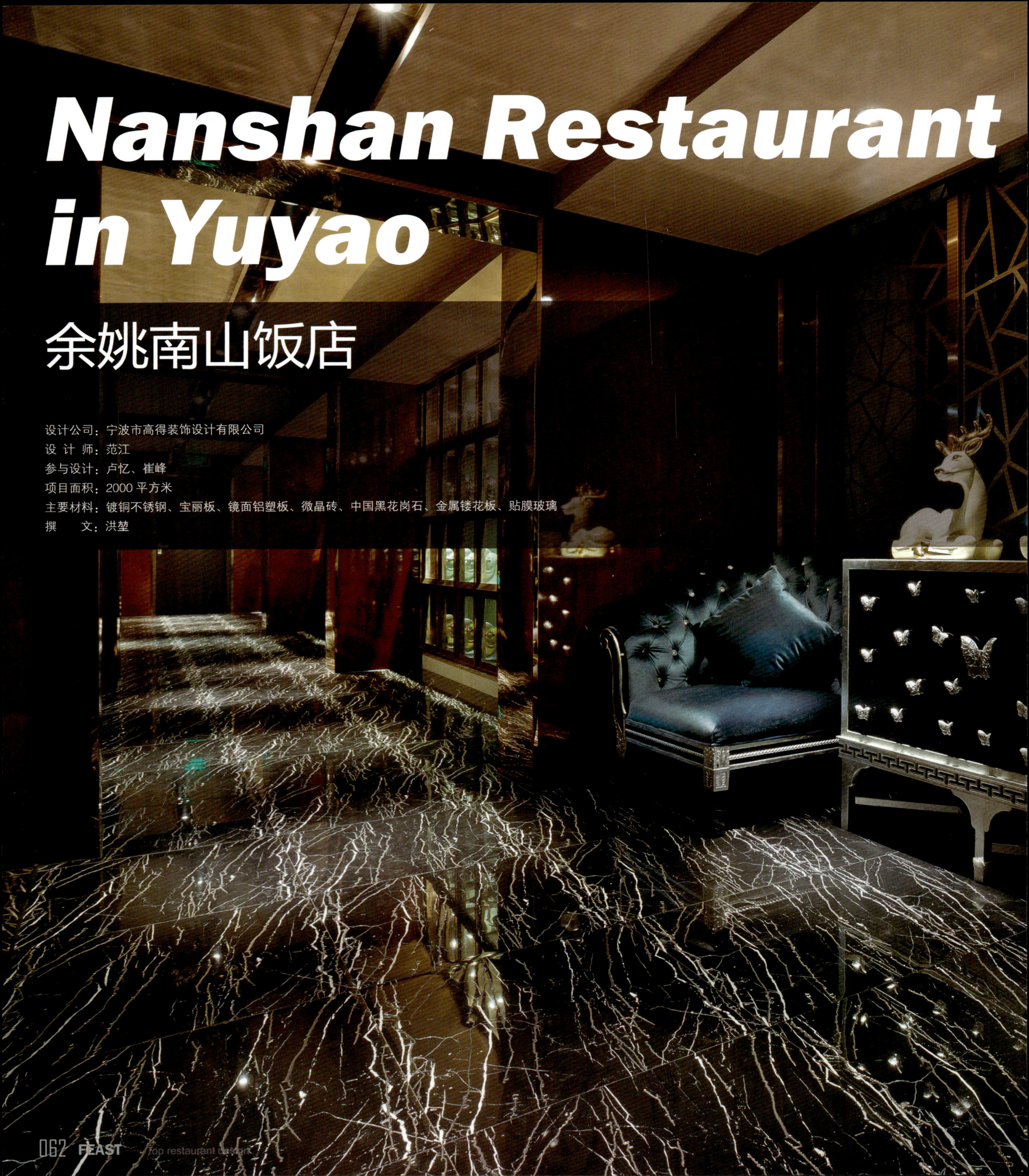

Nanshan Restaurant serves mainly for wedding banquet and birthday party, so the designer aims to create a space that combines old elegance with modern simplicity and shiny luxury. As the first floor is small, and the third floor is large, the designer reversed the tradition. The result is that the first floor serves as receiving area and ordering area, the second floor provides compartments and small dining tables, the third floor serves as banquet hall. As for the colors, red is widely used in the screen, glass partition, mirror, carving board and tables, which means auspicious and splendid. The furniture, including lightings adopts Neo-classical style. The tables and chairs are with bending lines. The color of light purple, black and white straight patterns and peacock blue are soft, making people feel elegant and comfortable.

The top surface of the first floor is made of wooden lattice mirror aluminum composite panel. There is a group of red and golden screen with Yuyao old scenery. The large ceramic square seal is engraved with the 100 common Chinese Family Names, which goes from the first floor to the third.

Entering the second floor, the Chinese style courtyard design creates a broad vision for people. The coin-made lattice are intercrossed, forming circular half-open pavilion; the hanging droplight's light is reflected in the aluminum composite panel, as if there's another lamp on the top. On the roof, various kinds of china vases are set around, so the light becomes blur when shine out, and in such an elegant environment, one couldn't feel it's such a precious occasion with such beautiful ambiance.

As for the design of the third floor, the slope has 4.8 meters to the roof. The suspended stainless glass board is connected like watch chain. The lightings seem like spreading fireworks. The soft furnishing is light green with golden pearly shiny fabric. By close look, the dark grey and white terrace is also like a watch chain, revealing splendid feeling, warm and hopeful, refreshing people who come for birthday party or wedding banquets.

南山饭店的经营主流是婚宴与寿宴，为此设计师把古朴的雅与现代的简洁和闪耀的奢华感作为本案的定位。由于一层面积小，第三层面积较大且有层高上的优势，设计师颠覆了原来的布局，把一层作为迎客厅与点菜区，二层为包厢和散座，三层为宴客厅。色彩上本案较多用到了红色，如屏风、玻璃隔断、镜子、雕花板、桌子等等，都寓意喜庆与辉煌。包厢门牌用的是中国各个朝代的年号，手书在半边瓷瓶上，内装灯光置于门边，包厢内梅兰竹菊的剪影，随意可见中国文人式的幽思。家具与灯具却运用新古典风格，曲线型的桌子和椅子，浅紫、黑白直纹、孔雀蓝等不同颜色的绒面，色彩柔和，让人感觉华贵舒适。

层顶面是木方格内嵌镜面铝塑板，接待桌后面是一组雕有古余姚风情图的红漆填金屏风。用陶瓷烧制的巨型正方形印章，上刻百家姓，凹凸有致，镶嵌在墙上贯穿于一至三层。步入第二层，有如中国庭院的设计让人视线豁然开朗，由铜钱变幻出的花格花纹，环环相扣，组成一个个圆弧形半开放的亭子；垂下的吊灯映在铝塑板镜面上，仿佛顶上也有一盏灯为你照亮。顶上骨瓷的花瓶灯形状各异，参差不齐，密密麻麻地环绕一周，透出的灯光朦朦胧胧，雅致间想起那一句：良辰美景奈何天！

在第三层的设计上，斜坡顶面，挑高 4.8 米，吊顶的不锈钢镜面板以手表链带的方式连接。灯具如散开的烟花，软包为淡绿色泛着金属光泽的珠光布，深灰与白色相拼的地坪细看也是链带的形状，流露出崭新夺目的气势，热忱而充满希望，用来排场庆寿及婚宴让人顿觉精神焕发。

Splendid Jinshan

锦绣金山

设计公司：北京成就辉煌室内设计顾问事务所
设 计 师：程辉
项目地点：石家庄
项目面积：4700 平方米
主要材料：奥金米黄石材、杭灰石材、黑白根石材、干黏石、实木雕刻、实木地板、壁纸

The stylist of this case is mainly based on the European elements, and it added some Chinese style elements, which can not only reveal the noble atmosphere of the space, but also show an attractive glamour in details. The design of each compartment is of different features with the different color tones and decorative items, which brings the whole space a sense of diversification.

For the choice of materials in hard outfit, the stylist uses marble, artificial sand rock, wallpaper and high-grade timber structural space, strengthening indoor atmosphere.

On display, the paintings are full of artistic aesthetic feeling combined with various kinds of furniture cooperating with the Chinese style screen and artistic carve patterns on woodwork which integrates the orient meticulous with elegance and romance in the western. The visual feasts loaded with oriental and western culture is put on and on…

In addition, the use of lamplight is the highlight of this case, and different types of lighting foils indoor objects with rich expressions building a kind of elegant and romantic dining atmosphere.

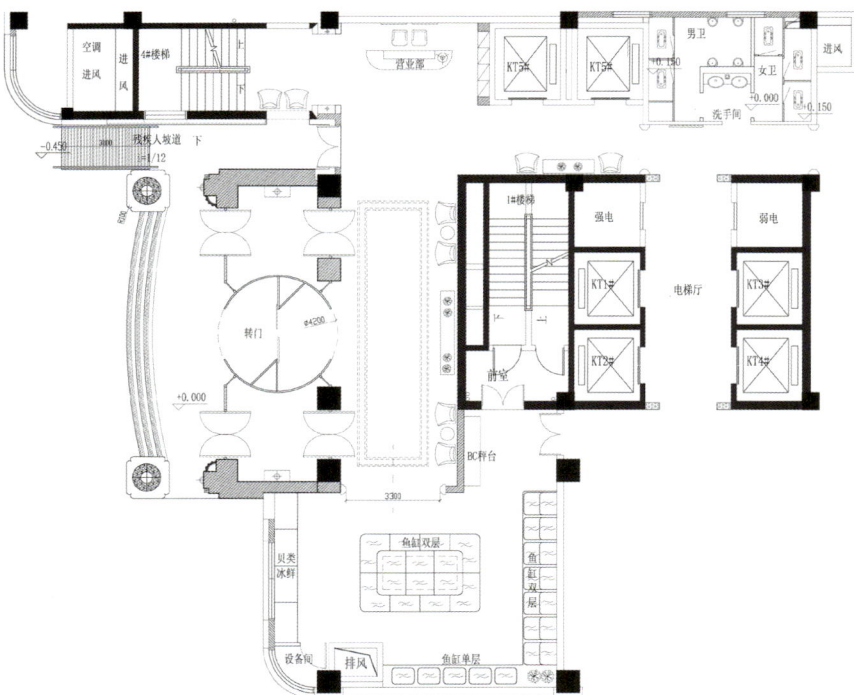

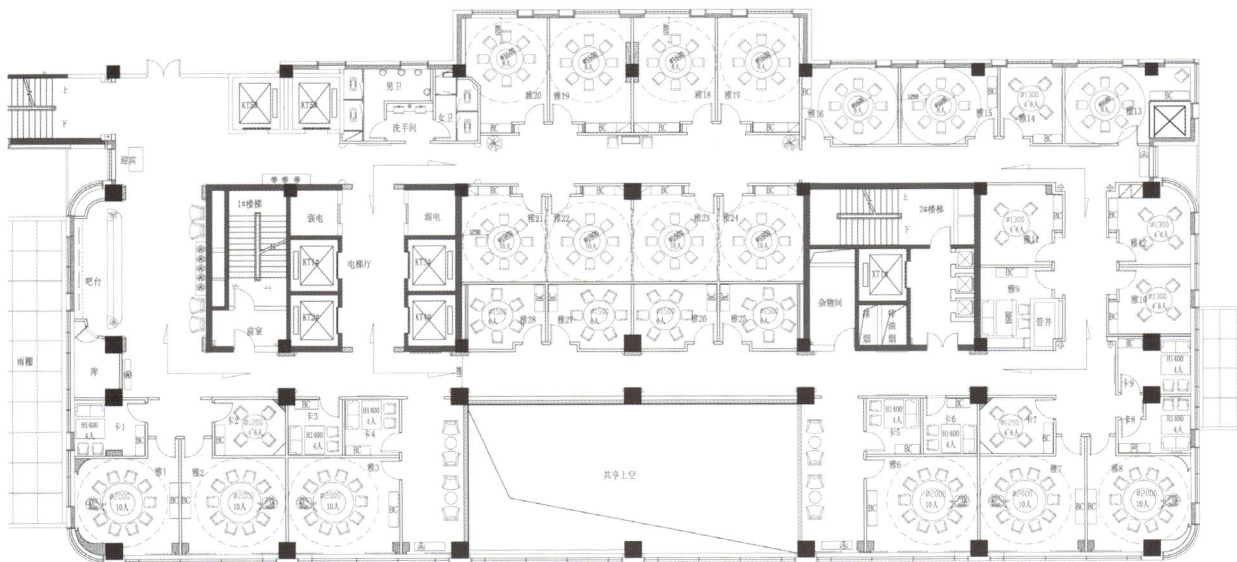

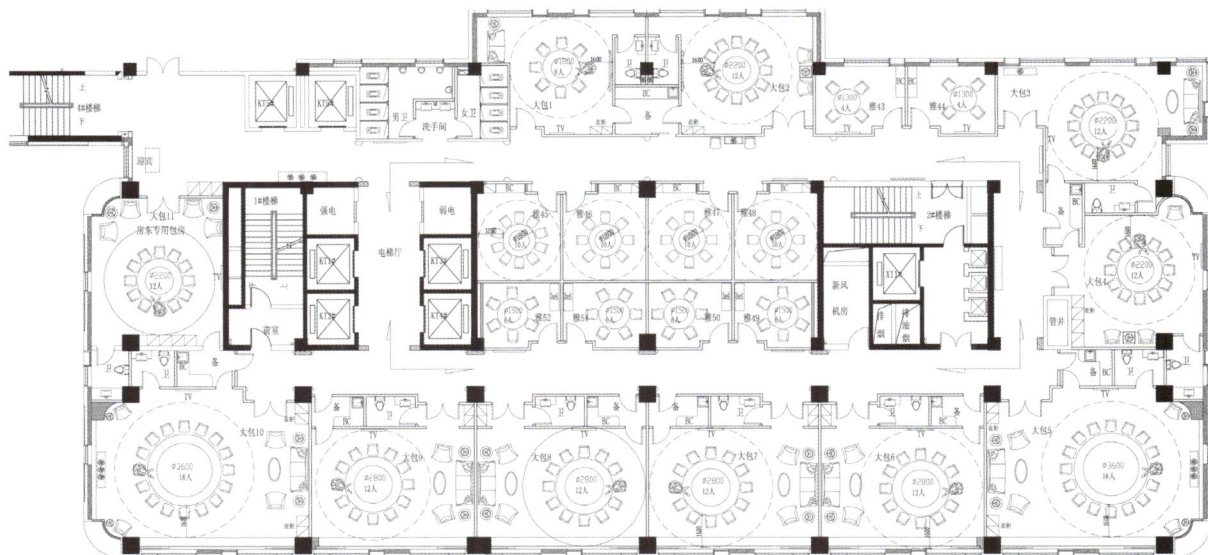

Jinghe Mansion

井河公馆

设计公司：北京大石代设计咨询有限公司
设 计 师：吴晓温、张迎军
项目地点：天津市

In this case, the landscape design adopts "serene", "elegant", "interesting" concepts to create an easy space, so that the customers can experience the pleasure of "visiting, studying, and appreciating" the garden. The green hill and lotus leaves at the entrance, the poems and verses in the lobby, the carps playing in the water together outline the comfortable and elegant life in Jiangnan.

Modern refined elegance is what the space presents. On the base of mortar walls and black tiles, the designer incorporates modern fashionable elements. The good use of colors make the space more cozy and contemporary. As for the furniture, simple modern style ones are chosen instead of the traditional officer cap chair. The color is correspondent with the fine and elegant space. As a result, the customers can enjoy the unique environment when dining here.

井河公馆 私家菜

井河餐饮
JINGHEGUANDI
RESTAURANT

RIVER RESTAURANT

本案的景观布置上采用"静""雅""趣"的理念，来打造出一个生活化的空间主题，使得客人在行走过程中体验"游园——读园——赏园"的空间情趣。门厅的青山碧荷、过厅的诗词歌赋、大厅的锦鲤戏水等共同勾勒出江南富雅的生活。

现代的儒雅是本案展现出的空间表情。设计师在提取粉墙黛瓦的色调基础上，加入现代时尚的元素，无色彩等系列色的运用使得清雅的空间多了份亲切感与现代感。家具没有采用传统的官帽椅而是简约的现代家具，从色彩之间的关系上让空间的造型与儒雅情节相呼应使得传统与现代形成一种情感的交融，让客人在此空间就餐别有一番感受。

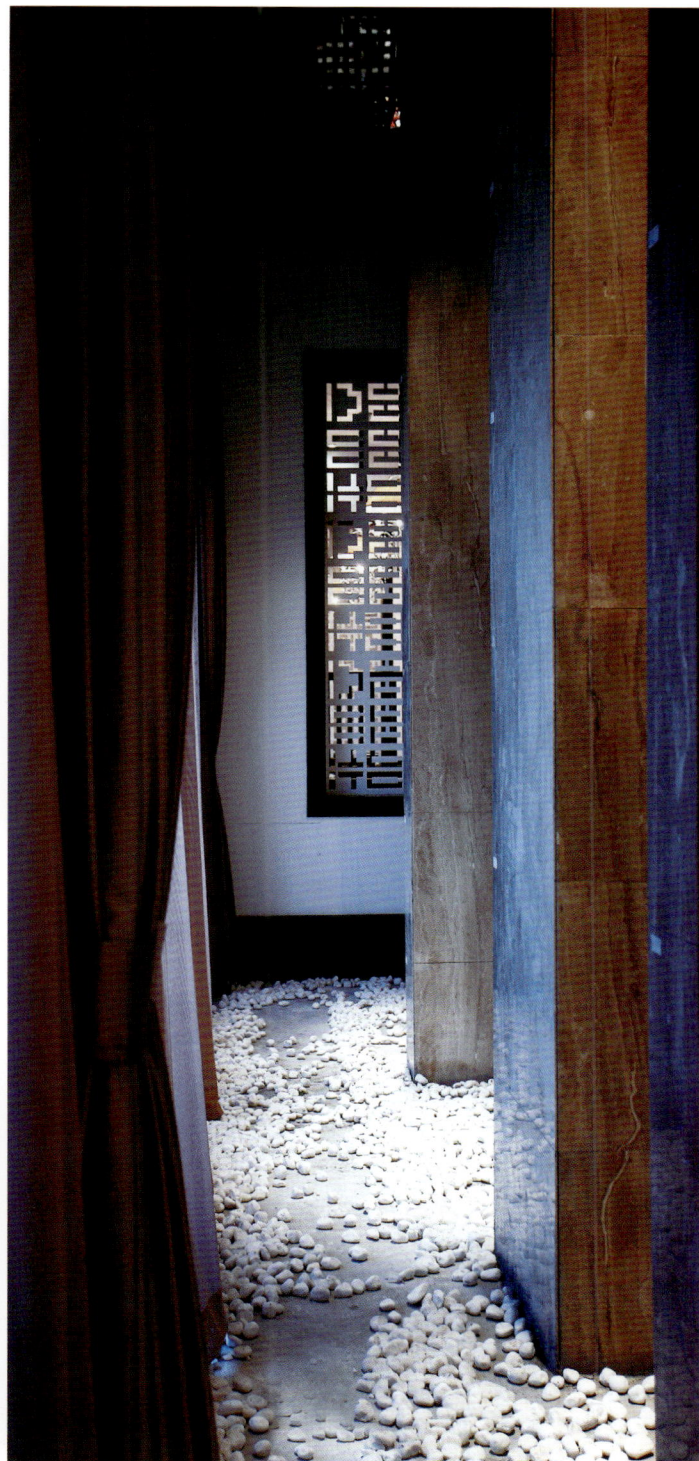

Huanghua Boyang Hotel

黄骅泊阳会馆

设计公司：北京大石代设计咨询有限公司
设 计 师：霍庆涛、丁洁
参与设计：汤善盛、赵洪程
项目地点：河北省黄骅市
项目面积：4500 平方米
主要材料：白木纹石材、白影木饰面、钨钢、素色壁纸

Huanghua, located in the north, is a sea port city around the flourishing Bohai bay. No matter the dietary structure or culture, it is closely connected with the sea. The whole building includes five floors. The first and second floor serves as a high-grade hotpot restaurant, the rest several layers are guestrooms. The owner takes the name "Boyang" also means moored in the east of sunrise, meaning their business to be flourishing.

Water has a variety of pattern, either still, or moving. Here the "water" form is fixed in various forms. The translucent water droplets sloping down from the top become solidification and photoed by the instantaneous image and then into a glittering and translucent "wall" – light romantic droplets lamp pool.

The wide use of living water landscape adds a lot of Reiki to the space. The falling waterfall walls, winding fold shui, thriving in the fountain decorate the space and also improve the environment for the north dry indoor environment. The use of light-colored wood finishes and grain, endows with the use of stone overall environment a bright and warmth tone, gentle and natural wooden texture, matched with the transparent waterdrops, forming a beautiful picture of "the land of woods and water".

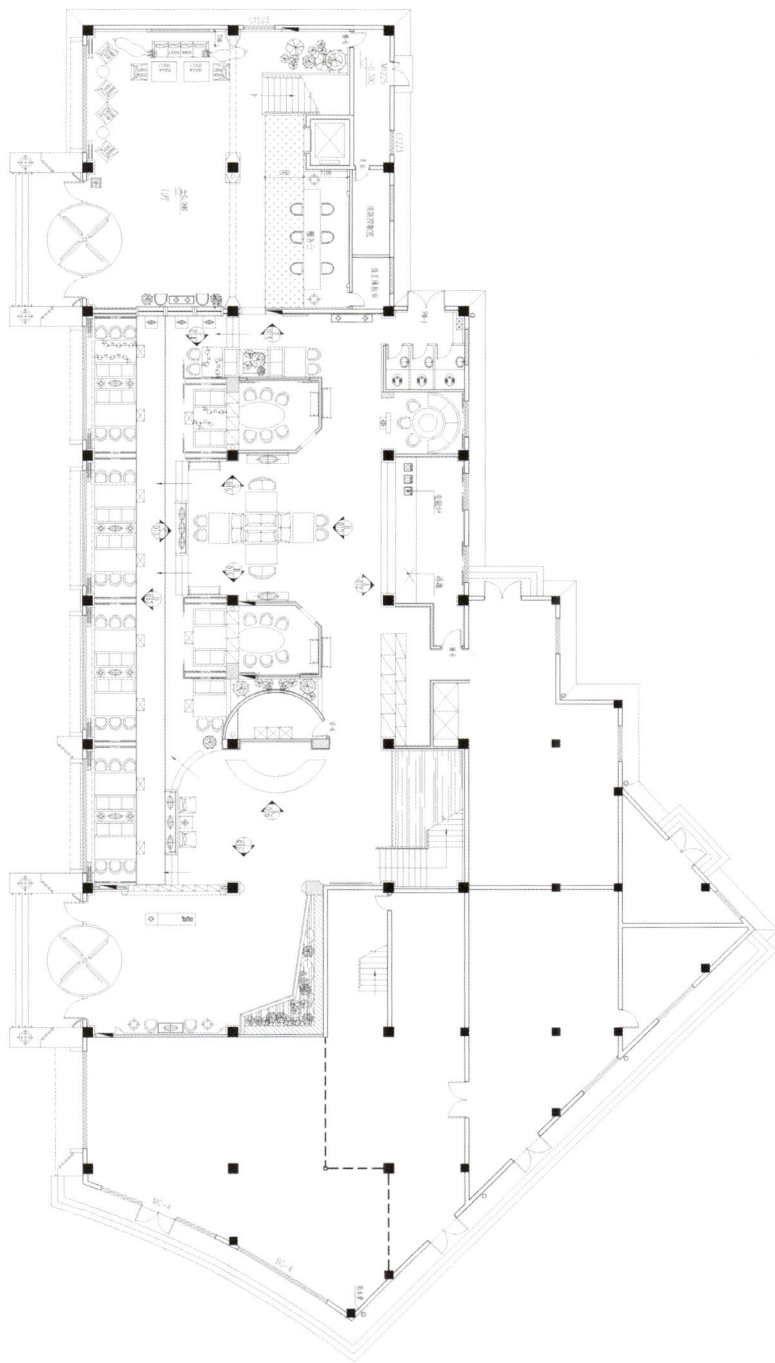

黄 骅地处北方，属新兴环渤海港口城市，无论从饮食结构还是文
化上都与海有着密切的关系。整幢建筑共有五层，一、二层是
一个高档的火锅餐厅，其余几层是客房。业主取"泊阳"也意指停
泊在东方太阳升起的地方，寓意自己事业蒸蒸日上。
水有多种形态，或静，或动。在这里，"水"的形态被空间凝固成
了多种形态，晶莹剔透的水滴从顶部倾斜而下，瞬间凝固，被捕捉
成影像，于是便有了那晶莹剔透的"墙"——光影浪漫的水滴灯池。
大量活水景观的运用也为空间增添了许多灵气，倾斜而下的瀑布墙、

蜿蜒曲折的叠水、蓬勃而出的喷泉，在为空间提供装饰的同时，也
对北方干燥的室内环境起到了一个很好的调节作用。
浅色木饰面和木纹石材的运用，赋予了整体环境一种明亮和温馨的色
调，柔美的天然木纹，与晶莹的水滴交相辉映，形成了一幅"水木清华"
的画面。

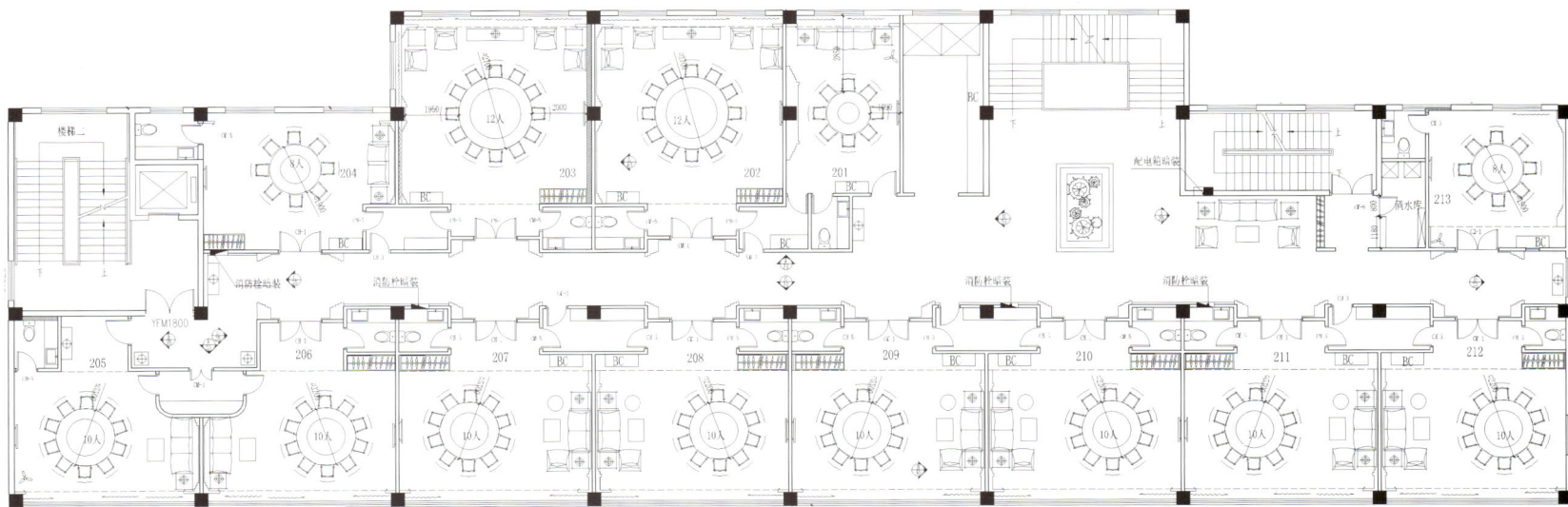

东方风情

Oriental
Style

First-class Flourish Yue Cuisine Mansion

一品和兴粤菜公馆

设计公司：湛江柏丽装饰设计工程有限公司
设 计 师：李宏图
项目地点：海口市秀英广场
项目面积：约 5000 平方米
主要材料：黑木纹、灰麻大理石、铁刀木

In this project, the designer uses very concise method to express Chinese elements, which is true and not exaggerating. But it is a little extravagant in the layout. On the first floor, there only places a reception desk and a small showing area. The hall adopts courtyard design to create patio gardening. The lights are divided into two kinds to create the effect of day and night.

One unique design is that the wall of each compartment is backward 2 meters, and a screen is formed between the pillars, creating a passage, which is across with the main passage, producing impressive space. The narrowest parts of the third and fourth floor both exceed 3 meters, and each compartment is recessed 2.7 meters to form a door area, which appears simple and clear. As for the decorations, the designer asked craftsman to copy the 12 animal heads statue to be the decorations of the third and fourth floor.

宴

本案设计上，设计师运用了非常简练的手法来演绎中国元素，化繁为简，写实低调。在布局上反而有些奢侈，一层工作功能只设置服务前台和一个很小的产品展示区；大堂引用天井手法设计了中庭园艺，灯光方面分为两组以营造出白天和夜晚的效果。二层设计上的精彩之处是每间包房原有的墙体再退缩两米，利用柱与柱之间做了一道屏风，形成了一条内过廊，与主通道之间虚实相交，具有强烈的空间感。三、四层走廊最窄的部分都超过三米，每个包房的门口都退缩2.7米，形成一个很有体量感的门关，力求简练。在陈设方面设计师专门找来工匠按1:1比例复制了十二兽首作为三、四层走廊的摆设。

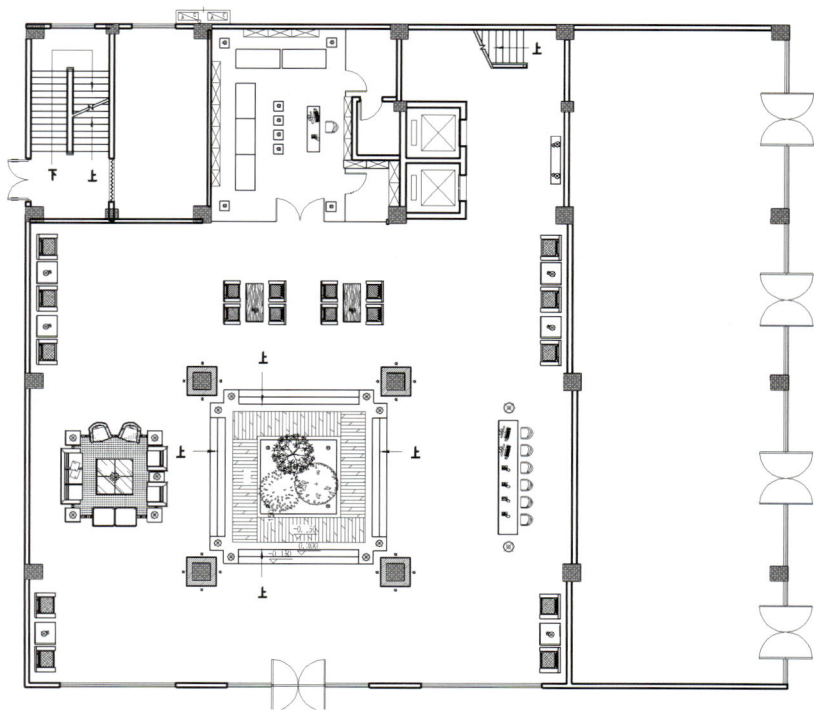

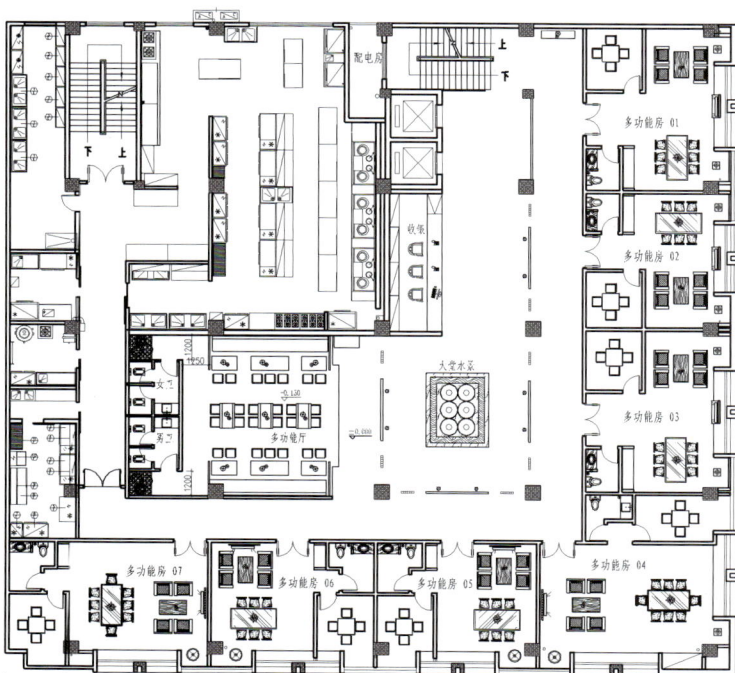

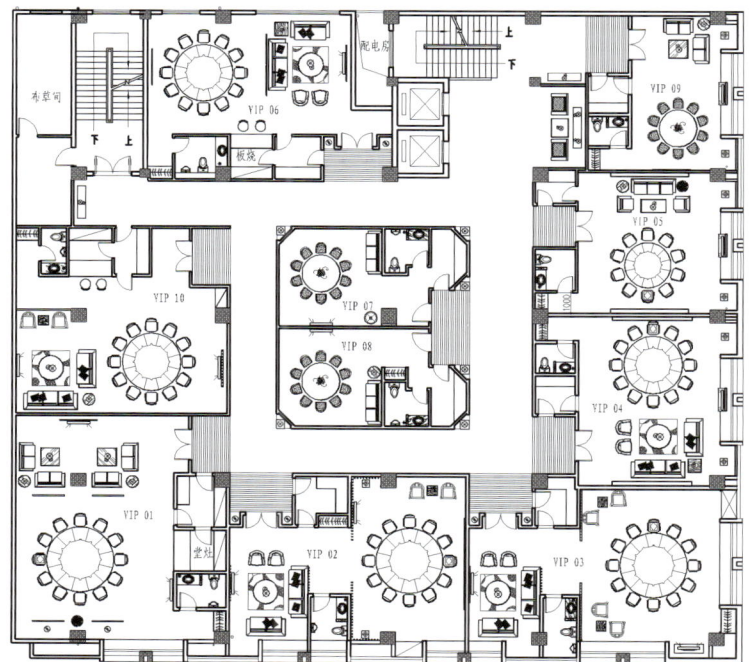

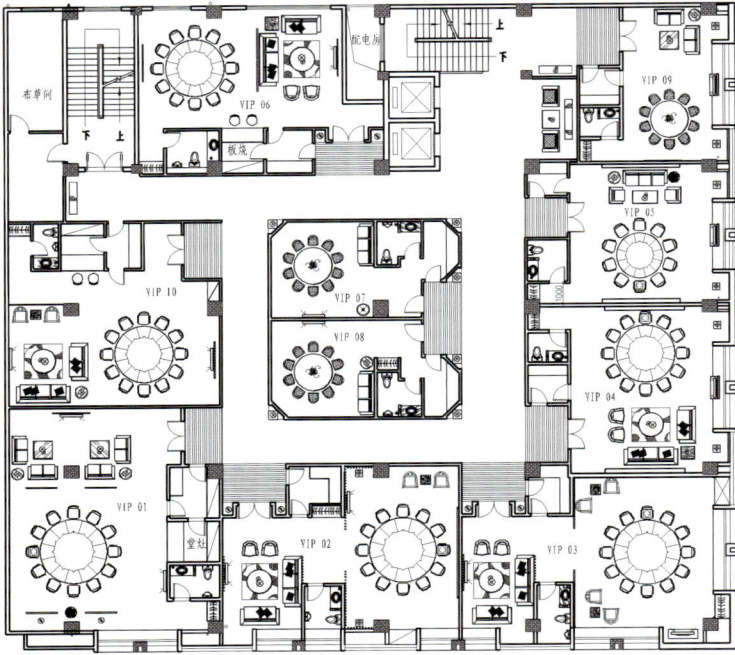

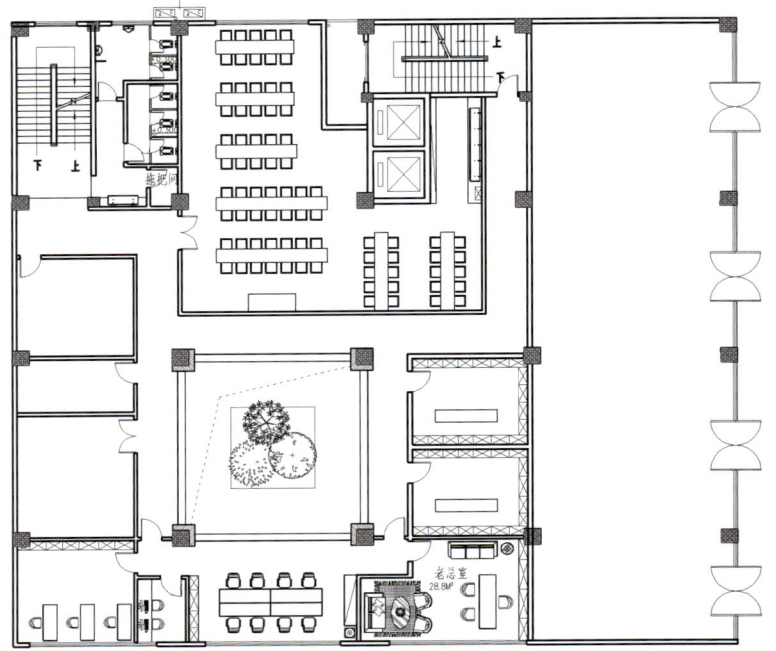

宴——顶级餐饮空间设计

Sherwood Landscape

西华山水

设计公司：昆明中策装饰有限公司
设 计 师：沈静
项目地点：昆明西华公园兰圃原址
项目面积：4800 平方米
主要材料：灰砖、木质、石雕、苏绣

SILVER LANDSCAPE RESTAUR

Hidden in the bustling urban, clear and refined environment edifying one's soul, these are the charming points of this case. The designer breaks Chinese elements into pieces and then integrates these into the space. All clear walls and all windows build the main wall of this space, to spread the primitive simplicity atmosphere in the indoor. The indoor furnishings foiled by the lamplight show verve and taste, calligraphies and paintings highlighting Chinese culture, various elegant and clear lamp, graceful and refined furniture from Ming and Qing dynasty, lifelike stone carvings, etc, which brings people's thought to the ancient times. In addition to building classical atmosphere, the stylist integrates the spacious indoor structure with the decorative materials which can highlight its imposing feature, building a noble and elegant large-scale public space.

Outdoor landscape is equally fascinating. The designer extends the nostalgic feeling to the outside, and offers the visitors a peaceful mind and a place to detach themselves from the bustling earth, among grasses, flowers, trees and waterside pavilion.

隐 匿于纷繁熙攘的大都市之中，用清浅雅致的环境陶冶人的心灵，是本案的魅力所在。

本案的设计师将中式元素打碎后融入到空间中。一面面清水墙、一扇扇格窗，构筑出空间的主墙，让古朴悠远的气息在室内蔓延开来。室内的陈设在灯光的烘托下更显神韵，彰显中华文化的书法、画卷，清丽雅致的各种灯饰，细致婉约的明清家具，栩栩如生的石雕……将人的思绪带到了久远的岁月之中。除了营造古典氛围之外，设计师还将

宽敞的室内格局与突显气势的装饰材料相结合，打造出一个高贵典雅且规模宏大的公共空间。

室外的造景同样精彩，设计师将怀旧的情愫延伸到了室外，让来访者在亭台水榭、花草树木之间找寻到宁静以致远的淡泊心境，超脱于尘世之外。

Mingyuan Cafeteria

名苑自助餐厅

设计公司：沈阳大展装饰设计顾问有限公司
设 计 师：孙志刚
项目地点：沈阳市
项目面积：1000 平方米
主要材料：法帝米黄理石、热转韧条、琉璃、金属马赛克

This case is located on the second floor of a leisure hotel. The stylist skillfully puts oriental culture inside it by using materials, in which the counter space presents a scene of literati pleasures, and all kinds of open-mode kitchens are more like a stage, showing Chinese culinary art beauty. In the dining area, scattered bamboo shadows, screen, the stone pillars and fresh bamboo, thousand fold with each other, act in cooperation with each other across a distance, and bring a thick wash-painting sense for this elegant picture.

The project makes the lychee-surface stone as the major interior decorative material, building a building's special aesthetic taste and function. The whole space uses the oriental culture as the basic element. There are different seat partitions, in which Chinese archaize cases, archaize food boxes and Chinese paintings of flowers and birds can be seen everywhere, which expresses the designer's skillful design of getting ancient and current ages as well as Chinese and Western style together. It creates an innovative and challenging artistic dining space by the modern and abstract technique with materials and colors designed and applied.

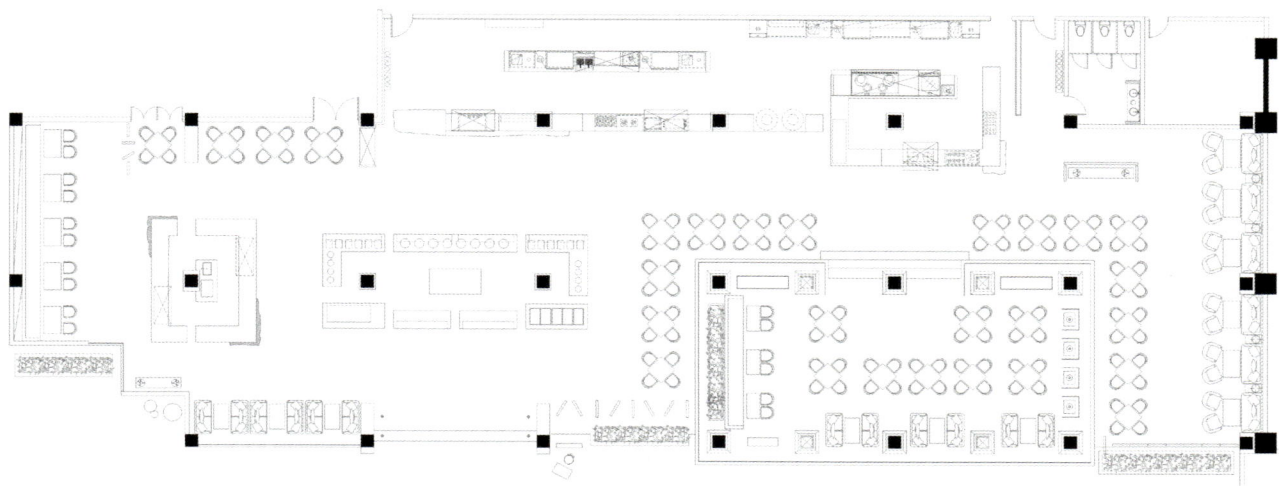

本案位于一休闲酒店的二层，设计师运用材料巧妙地将东方文化融合其中，散台空间呈现出一派文人宴乐的景象，各式开放式厨房更像一个舞台，展示着中国烹饪艺术之美。用餐区内竹影斑驳，屏风、石柱、翠竹，层层叠叠，疏离隔透，为清雅的画面带来浓浓的水墨之音。

该项目用荔枝面锈石作为内部装修主材，营造出建筑的特殊美感与功能。整个空间以东方文化为基本要素，在不同座位分区中随处可见的中式仿古案子、仿古食盒与中式工笔花鸟画等表达了设计师刻意交织古今与中西于一体的设计巧思。以现代、抽象之手法，加上材质、色彩的设计运用，为本案塑造一个创新而具艺术性的用餐空间。

Xunchang Story Restaurant

寻常故事餐厅

设计公司：福州创意未来装饰设计有限公司
设 计 师：郑杨辉
项目面积：550 平方米
主要材料：藤制品、仿古砖、装饰皮革软包

Every restaurant has a story, a theme. The theme of this project is the revival of the simple New-Oriental style natural environment. Luxury and fashion can be felt in the serenity. Cane products play a major role in the space. The materials are simple, matched with modern design techniques. Various cane lamps and lanterns express serenity of the space. In the warm light, people can appreciate modern aesthetic design while they enjoy the food to the full. The cane made cloud lantern remind people of the white cloud in the blue sky. The red yarn wall in the passage imbues warmness to the space, with the aim to let everybody remember the loving of the space, the unique characteristic of Xunchang Story Restaurant when they touch the soft red wall. The red lines on the unreal partition wall are the extension of color. The irregular European mirror frame and Chinese old carving are combined on the wall, which blur the boundaries of the space. The Chinese and European carving byobu matches with the various design elements in the space, showing the style of the space.

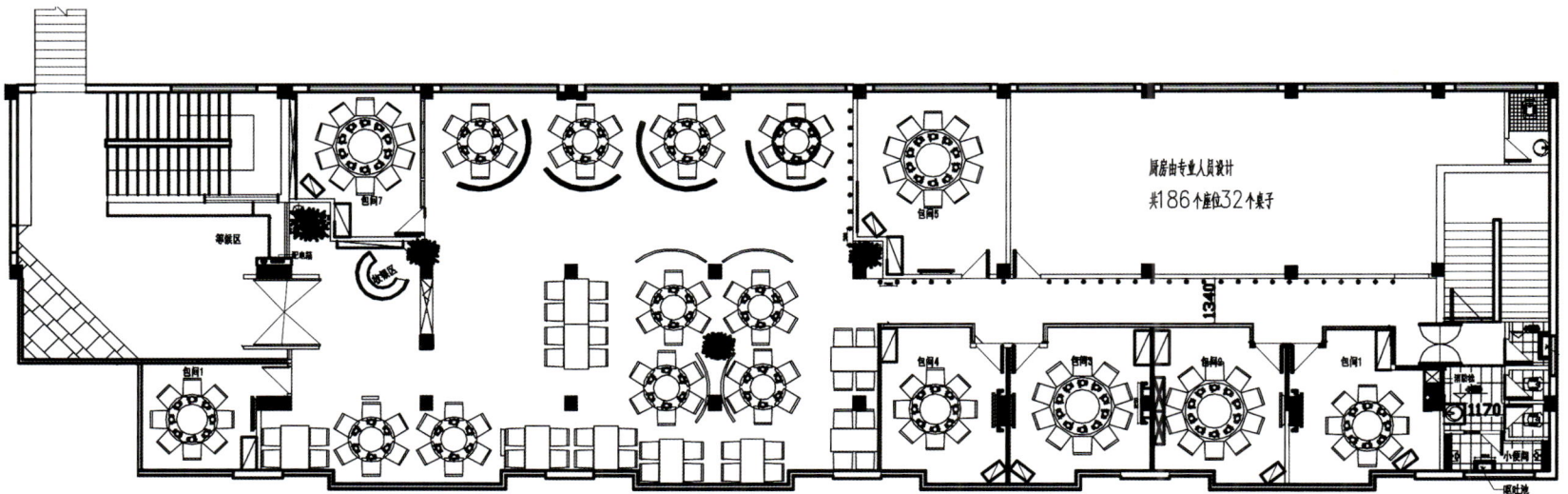

厨房由专业人员设计
共186个座位32个桌子

每间店铺都有一个故事，一个主题。本案的故事主题是质朴的新东方原生态空间场景的再现。宁静中的奢华时尚淡淡散发，空间场景营造的主角是藤制品。材质的选择是质朴的，设计手法造型是当代的，各种藤制造型定制的灯具在空间中诉说宁静质朴，暖色系的照明让置身于这样空间氛围中的食客的所有味觉都发挥到了极致，慢慢享用美餐的同时也有了当代视觉的审美熏染。藤制品的云彩装置吊灯造型，是意念中的蓝天白云，通道的红色纱墙温暖着整个空间，只愿每个来临寻常故事的都市人，抚摸红色纱墙柔软的质感触觉，就能记住餐厅的空间给予的无限温情，记住寻常故事不寻常的空间标识，虚隔断上红色的线绳是色彩的延伸，不规则欧式镜框和中式老雕花组合在空间的墙面上，模糊了空间的界限；中式和欧式雕花图案组合的屏风，和空间设计元素的混搭，也传达出空间想要诉说的。

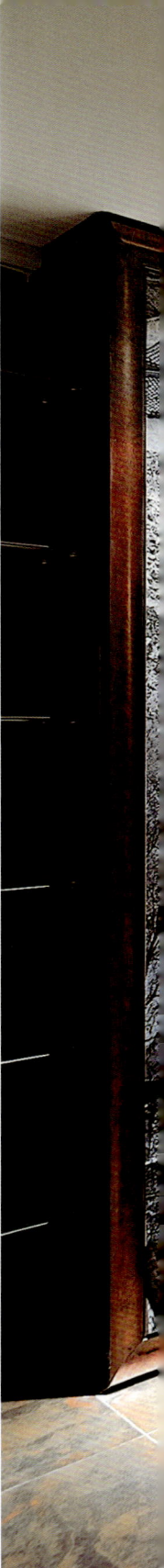

Fisherman's Wharf

渔人码头

设计公司：北京成就辉煌室内设计顾问事务所
设 计 师：程辉
项目地点：石家庄
项目面积：3500 平方米
主要材料：浅啡网石材、仿木纹砖、灰色仿古砖、木材、壁纸、仿铜制品

The dream for reviving a dynasty begins from this place...

The majestic Roman column goes throughout the whole space which offers people the king's imposing manner. Elegant and delicate lighting not only reveals the flavor of past ages, but also promotes the indoor style. The high-quality printing wallpaper decorates the wall surface for elegance and imposing manner. The ceiling design adopts the traditional wooden grid and plaster to build, manifesting the pure European style. Vivid and delicate embossment seems to pour out an old story, which becomes the focus of the corridor and has the effect of bringing out the crucial point. In order to get the best result, the stylist combines color with lighting, successfully building a kind of taste.

段复兴王朝的美梦从此处开始……

　　雄壮威严的罗马柱，贯穿整个空间，赋予其王者的气势；优雅精致的灯饰，不但流露出岁月的味道，而且提升了室内的格调；优质的印花壁纸，将墙面装点得典雅大方；天花板的设计采用了传统的木格栅和石膏来打造，体现出纯粹的欧式风格；生动细致的浮雕，仿佛在诉说着古老的故事，成为了过道处的焦点，有着画龙点睛的作用。

　　为达到最佳的华贵效果，设计师将色彩与灯光融合起来，成功地进行了情境营造。

Jap Cafeteria

日式自助餐厅

设计公司：Nota Design International pte Ltd
设 计 师：Keat Ong, Gary Zeng, Zhigang Sun
项目地点：沈阳清水湾休闲酒店二楼

For the Jap Cafeteria, Light stainless steel chains are utilized as drapes while suspended "rocks" (surface-treated concrete secured to stainless steel cable) act as physical screens. Light weight timber trellis interfaces with the dining spaces for a light ethnic touch. Japanese lacquered table tops act as subtle cultural reminder. Sitting on the other quieter side of the floor are the tea appreciation area, rest lounges and the massage rooms. The "floating" Tea Deck surrounded by feature pond acts as the area marker on the sprawling plan.

在本案中，轻型不锈钢的链子被用来当做布帘，悬挂的"岩壁"（表面处理过的把不锈钢电缆绑住的混凝土）也作为天然的纱窗。轻巧的木头凉亭格子把餐厅的空间结合得浑然天成，这是一次轻松的民族风接触。日本顶级真漆餐桌也巧妙地作为文化标志进行提示。 在楼层的另一边则更安静，有着品茶区、休息室和按摩室。"漂浮"的茶板被具有特色的池塘环绕着，这也延伸出本案的设计理念。

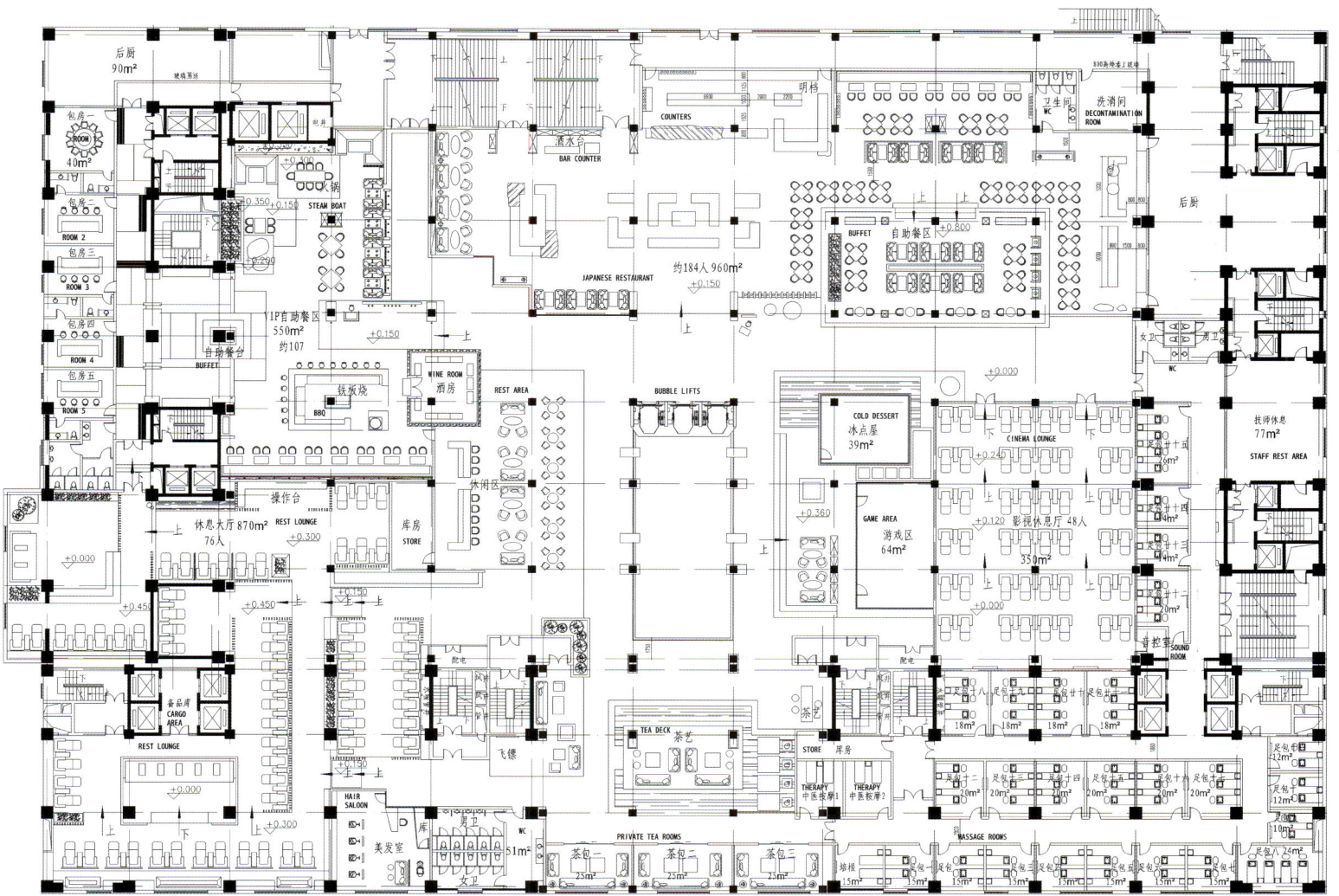

后厨 90m²
包房一 ROOM 40m²
包房二 ROOM 2
包房三 ROOM 3
包房四 ROOM 4
包房五 ROOM 5
玻璃幕墙
火锅
STEAM BOAT
+0.350 +0.150
+0.200
VIP自助餐区 550m²
约107
自助餐台 BUFFET
+0.150
铁板烧
BBQ
WINE ROOM 酒房
REST AREA
休闲区
操作台
休息大厅 870m² 76人 REST LOUNGE
+0.000
+0.450
+0.300
+0.150
库房 STORE
配电
备品库 CARGO AREA
REST LOUNGE
+0.000
+0.300
飞镖
HAIR SALOON
美发室
男卫 WC 女卫
配电
+0.150
+0.150

COUNTERS
酒水台 BAR COUNTER
JAPANESE RESTAURANT
约184人 960m²
+0.150
明档
BUFFET 自助餐区
+0.800
洗消间 DECONTAMINATION ROOM
卫生间 WC
后厨
女卫 男卫 WC

BUBBLE LIFTS
+0.000
COLD DESSERT 冰点屋 39m²
+0.360
GAME AREA 游戏区 64m²
+0.240
CINEMA LOUNGE
+0.120 影视休息厅 48人 350m²
监控室
SOUND ROOM
技师休息 77m² STAFF REST AREA
足包十一
足包十二
20m²
4m²
18m² 18m² 18m² 18m²
THERAPY 中医按摩1
THERAPY 中医按摩2
足包二 20m²
足包三 20m²
足包四 20m²
足包五 20m²
足包十六 20m²
足包八 24m²
TEA DECK 茶艺
茶艺
STORE 库房
PRIVATE TEA ROOMS
茶包二 25m²
茶包二 25m²
茶包二 25m²
足包一 15m²
足包二 15m²
足包三 15m²
足包四 15m²
MASSAGE ROOMS
12m²
12m²

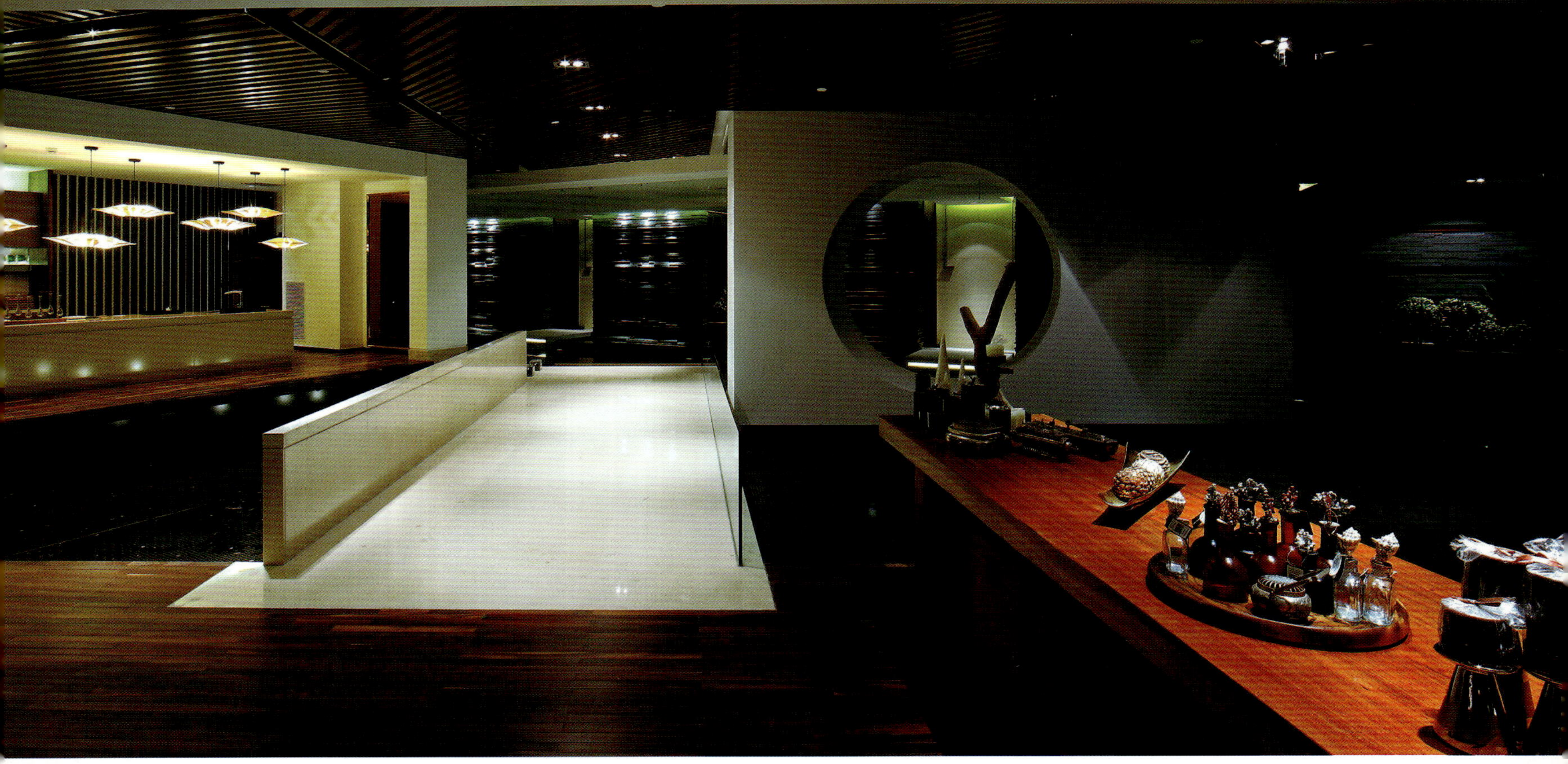

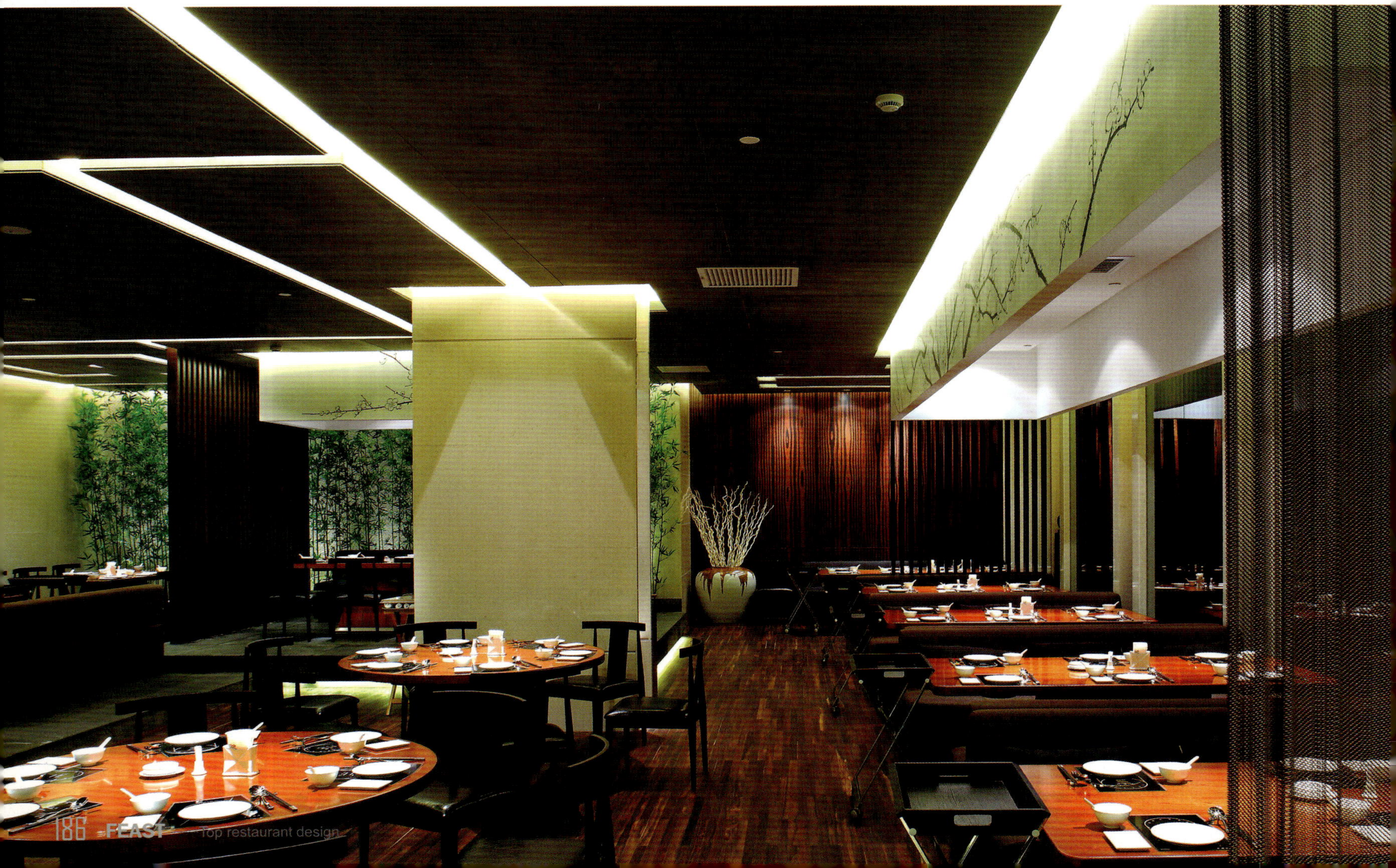

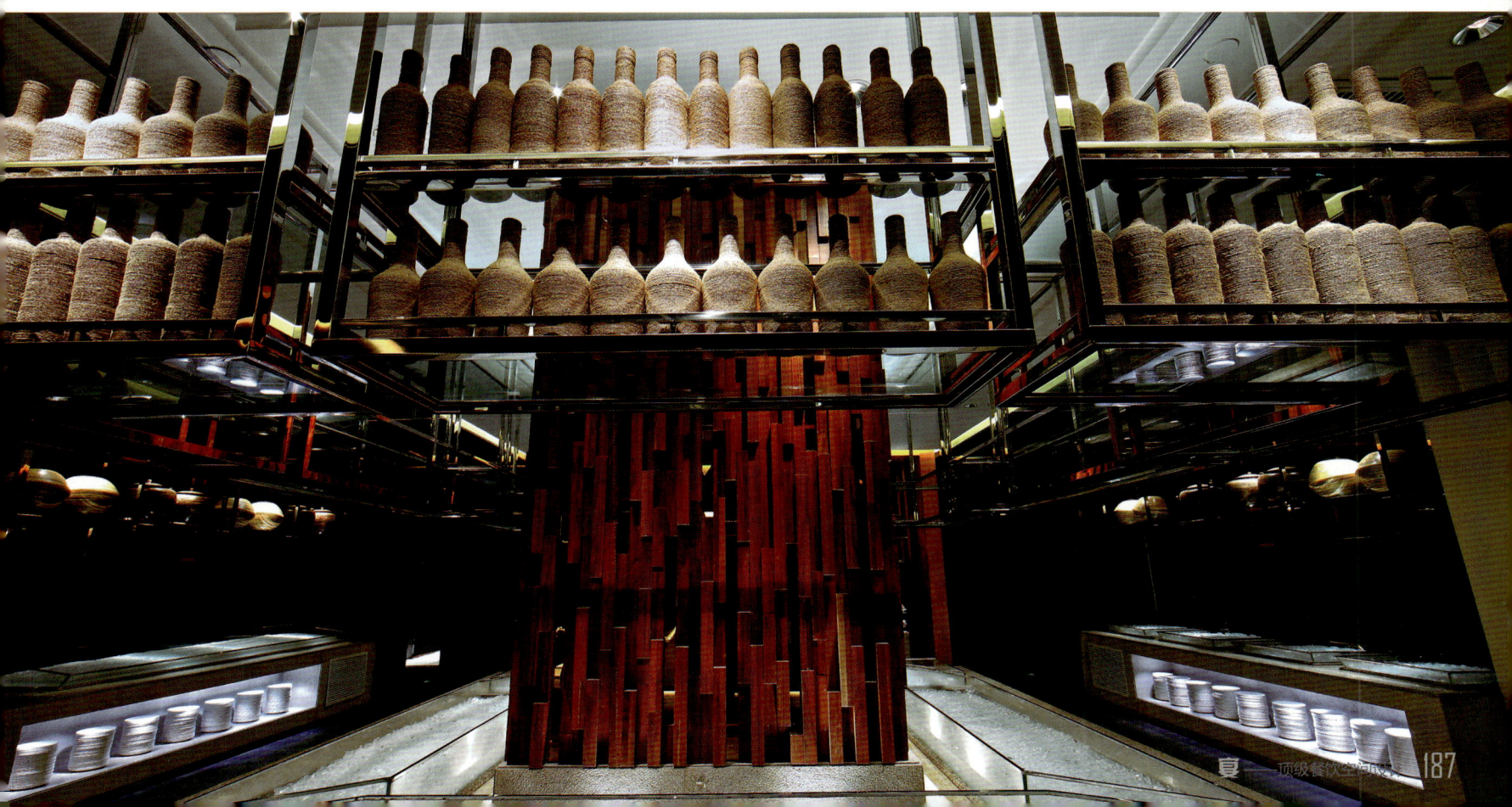

Dongguan Yuantong Sushi—Mingcheng Store

东莞元通回转寿司名城店

设计公司：广州文智设计工程有限公司
设 计 师：韦文生
项目地点：东莞
项目面积：1300 平方米
主要材料：桔黄石、柚木饰木、绘画感的地砖、茶镜

In this case, the design focuses on the façade, lobby, stairs and passage. In order to let the customers enjoy themselves while having dinner, the space aims to present a variable and subtle space ambiance.

As for the major materials, the designer chooses simple orange stone, teak and picturesque floor tiles, simple yet variable, different yet harmonious. The façade of the restaurant is a large French window. In the western side, colorful paper and yarn curtain dim the sunshine; at night, in the soft light, the space appears more romantic for dining.

Large mirrors are used in the stairs; in the light of the long hollowed wooden lamps, as if it's a three-dimension space. The customers can enjoy the fantastic visual feast while having dinner or walking, and they'll understand the rhythm of the space.

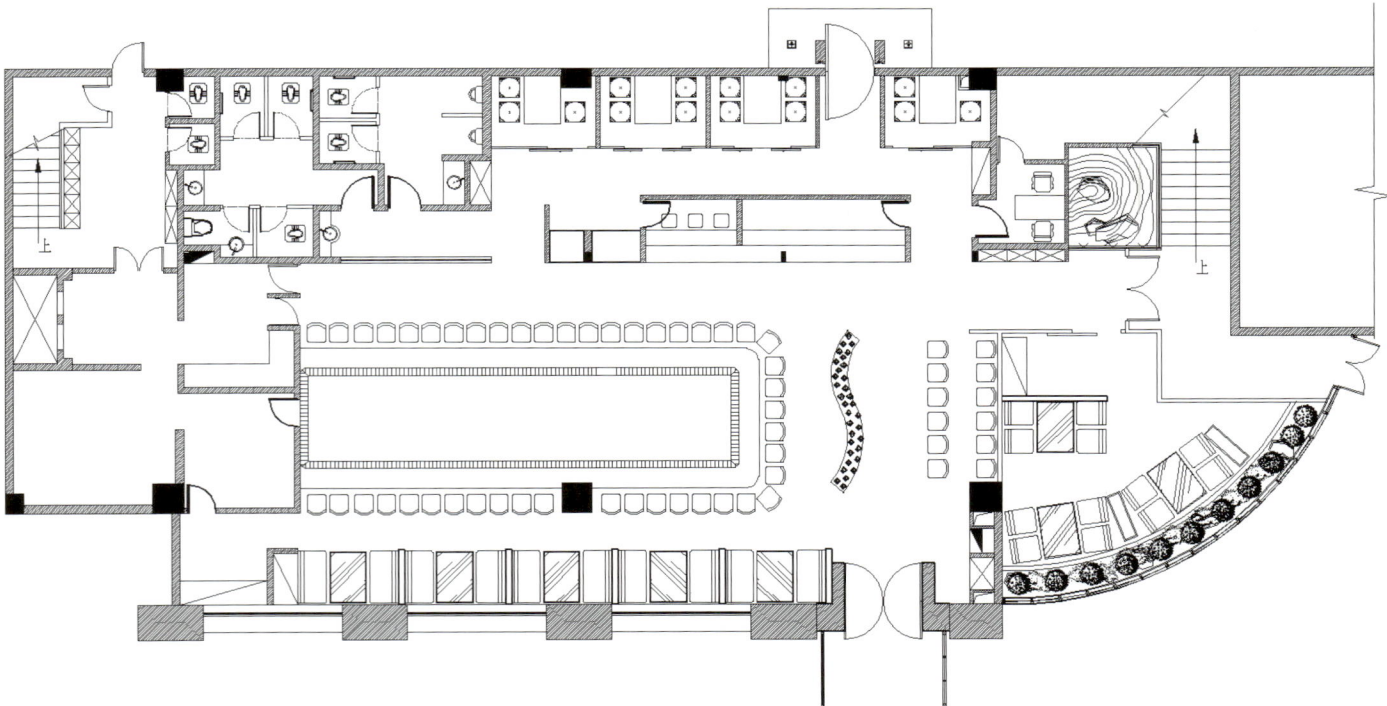

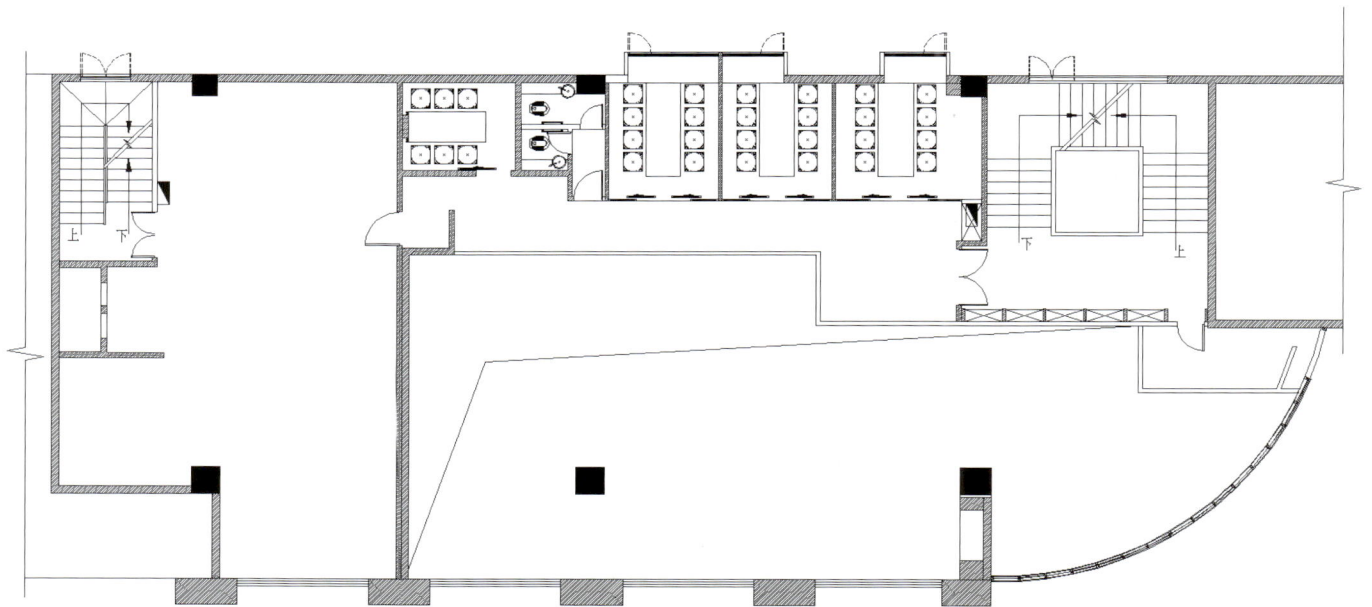

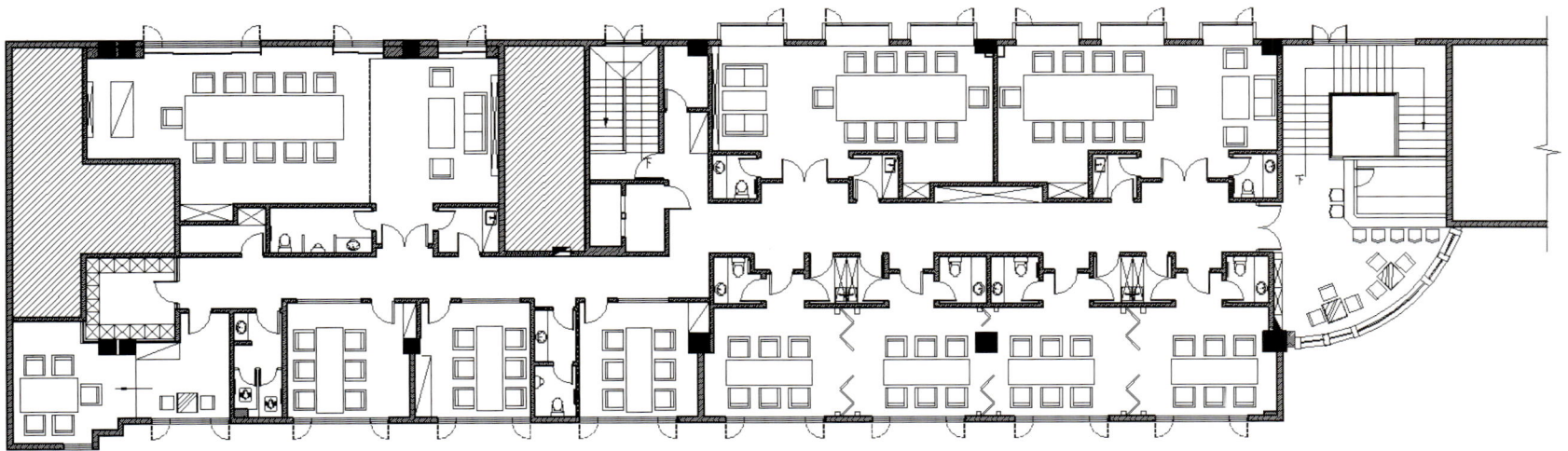

本案的设计重点在于外立面、大厅、楼梯、走道。空间注重视觉的深浅起伏，微妙的动静变化，从而让客人在享用美食的同时能有着喜悦的心境。

设计师在主要材质选用上，以朴实感的桔黄石、柚木饰木以及很有绘画感的地砖为主，简单中寻求变化，变化中寻求统一。餐厅的外立面是大面积的落地窗，西面通过彩纸、纱帘把强烈的光照控制减弱；夜晚，在柔和的灯光衬托下，更显浪漫，也更有用餐情调。

楼梯空间运用大型的镜面，在中空的长形木条饰灯照射下产生虚幻的三维视觉空间，客人在走动或享用美食时，享有奇妙的视觉感受，更容易贴近空间的旋律。

Dongguan Yuantong Sushi — Changqing Store

东莞元通回转寿司长青店

设计公司：广州文智装饰设计工程有限公司
设 计 师：韦文生
项目地点：东莞
项目面积：650 平方米
主要用材：地砖、柚木饰面板、墙纸、玻璃、茶镜

From functional layout, the restaurant is divided into two parts: the hall and the compartment areas. Considering Japanese culture and contemporary Chinese consumers' demand and habits as well as the architectural structure, the design creates a changeful consumption space. The walls all around the stair case are decorated with tawny mirror and different modelling heading wood panels, as if the mirror is in the space, yet the space was inside the mirror; meanwhile, decorations such as pots, make the stair space steady, elegant, and interesting. In addition, the Chinese traditional garden box scene from the landscape of the space and the leakage provide the guests different scenes when they move in the restaurant, which divides the spaec reasonably and forms some semi-enclosed seats.

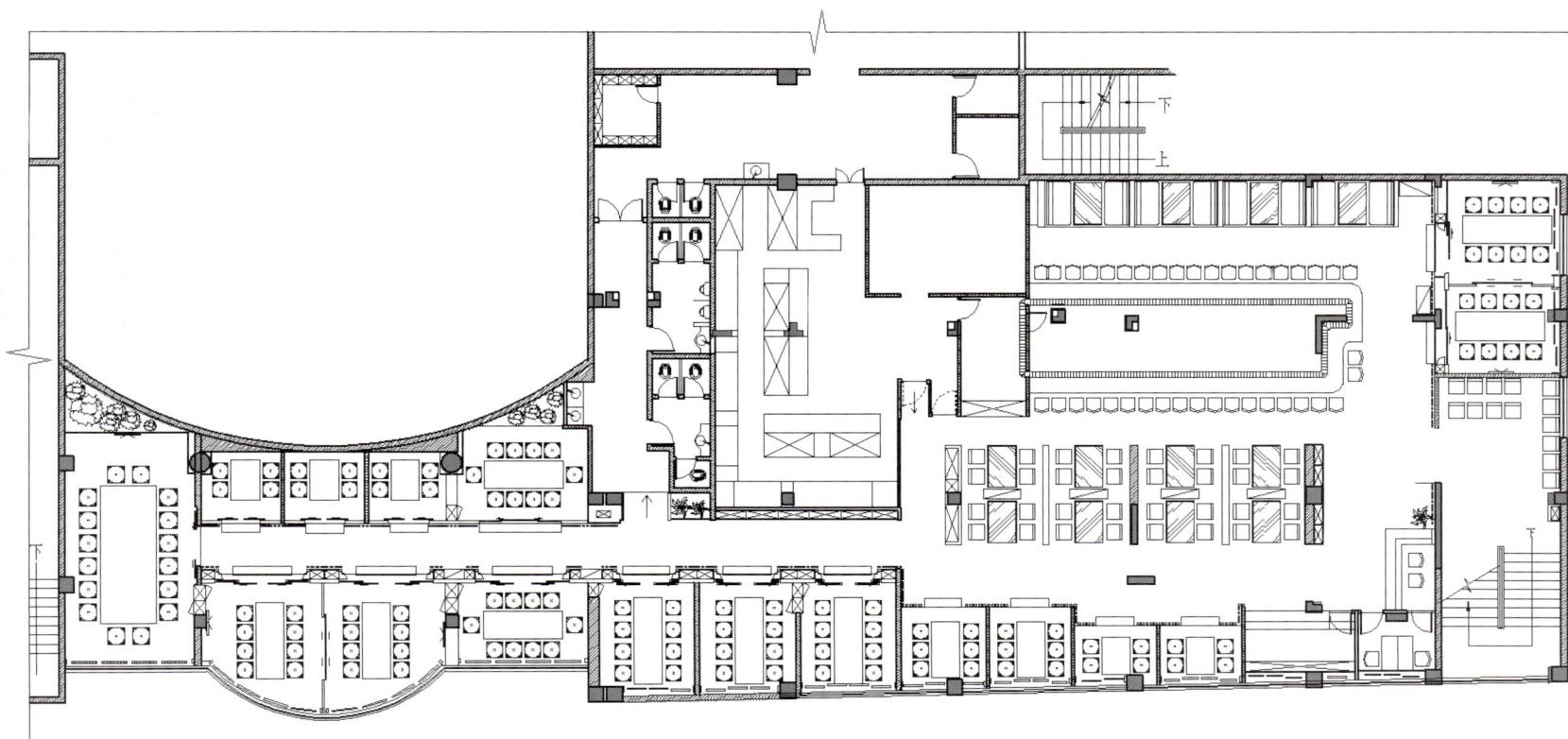

The compartments are enclosed space, elevated from the corridor by the entering slates and the thresholds. The design adopts a range of natural elements to organize the space, such as water ripple, tree bark, leaves, bamboo, and elements of Japanese culture and architectural culture. After deformation, refinement, and recombination, these elements become the important elements that connect the compartments and hall, and also the important step to create visual impact.

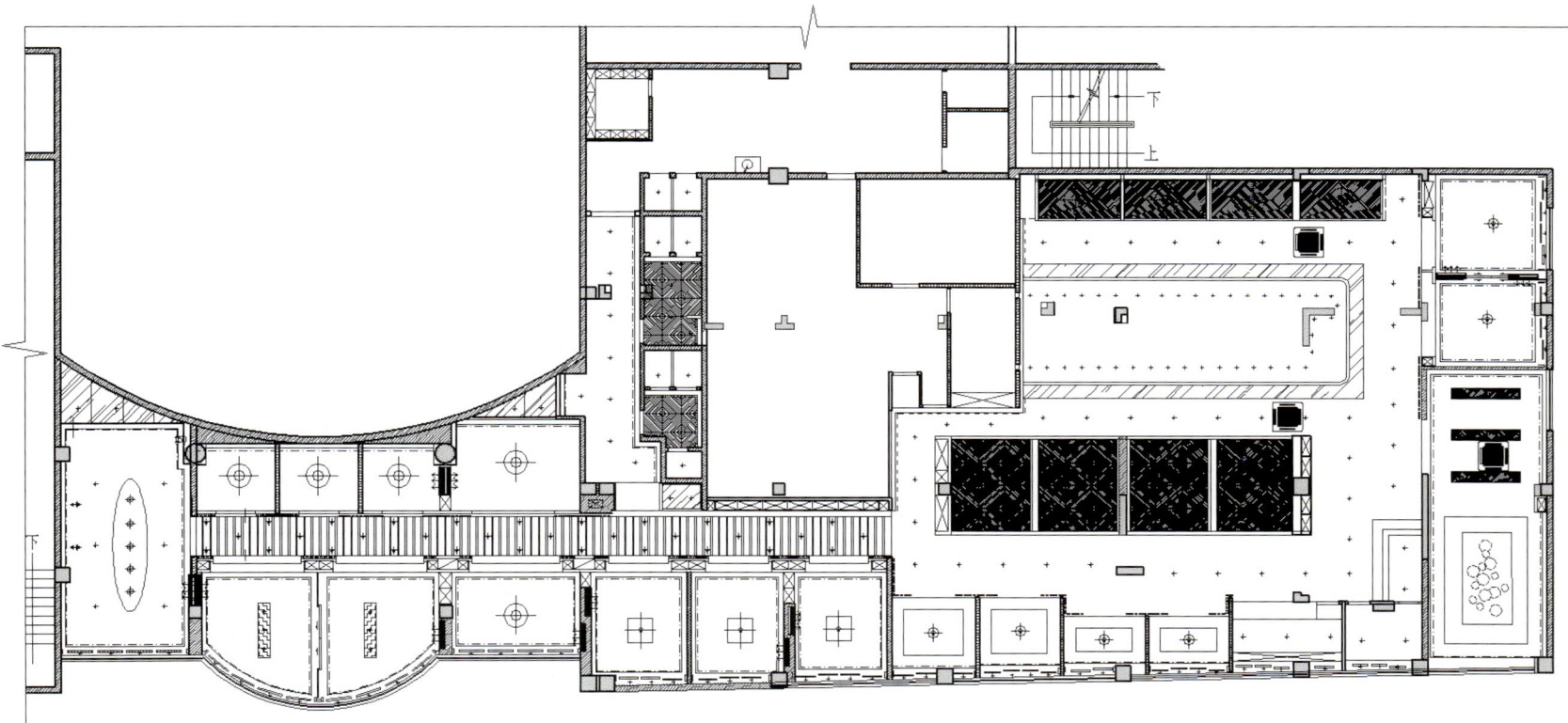

本案从功能布局上分为大厅、包间两大区域，运用日本文化意念结合中国当代消费者的需求与习惯，充分与建筑结构相结合，组合成一个多变丰富的消费空间。

梯间入口的墙壁四面用茶镜与不同造型朝向的木饰面板相结合；镜在空间内，空间在镜中，亦真亦幻；同时摆设陶罐等一系列的装饰物，使楼梯空间沉稳而不失雅趣。

另外，吸取中国传统园林框景、漏景的空间设计手法，通过客人的参与互动，达到移步异景的效果，从而合理地分隔空间，构成一个个半私密空间组合的卡座。

包房利用入口托石板、门槛与走道形成高差，从立面到平面分隔出私密空间。设计中采用了一系列来源于自然的造型元素来组织空间，如水波纹、树干、树叶、竹子与日本文化、建筑文化，经过变形和提炼后，再重新有机组合。这些元素是联系包房与大厅的要素，也是营造视觉冲击的重要步骤。

KAMIKAWA Japanese Cuisine Restaurant

上川日本料理

设计公司：徐庄设计（广州）有限公司
设 计 师：庄仙程
参与设计：周良娇、陈育谦、庄仙任、陈晓佳
项目地点：广州
项目面积：800 平方米
主要材料：人工自然石、实木、灯纸、荔枝面石材

In this restaurant, the space is divided into two areas, the cuisine area and the teppanyaki area.

The design for the space creates a simple serene environment. Entering from the right door of the hall, one can see the dining lobby, and be impressed by the 10-meter long sushi bar which leads to the natural stone teppanyaki bar. The teppanyaki bar and sushi bar become the DJ bar which plays a leading role allowing the customers to feel the enthusiasm.

On the left side of the hall is the teppanyaki compartment area, the solid wood table creates cozy and happy dining atmosphere.

It is as if a self-contained world, where the chef creates delicious food with pleasant appearance and smell, as if creating a new world in a moment attracting the customers. There's no need to worry that the joy and delicious food will slip away. People enjoy the happy time to the full here.

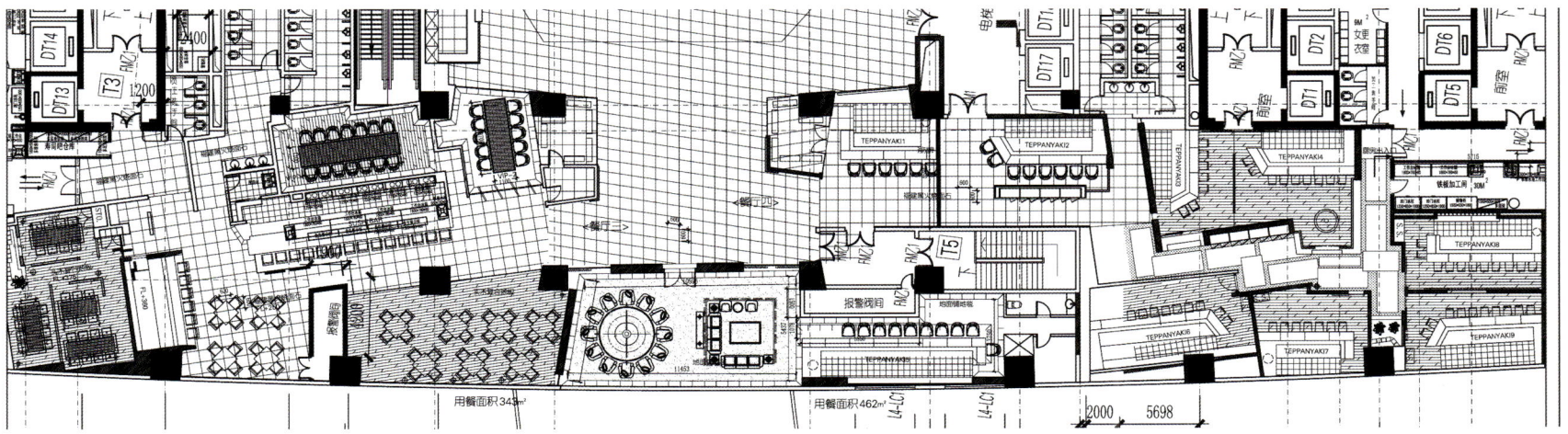

用餐面积345m²

用餐面积462m²

2000   5698

本案空间规划上分为两个区域，料理区及铁板烧区。

空间设计的整体就是给客人原始素朴的宁静空间环境，大厅右手边进去就是料理区的用餐大厅，远处第一视觉就是从十米长的寿司吧台带你到了自然石铁板烧吧台，从而铁板吧台及寿司吧台成为 DJ 台，起了主导的作用，使其他散座的客人也感受到了热情！

大厅左手边是铁板烧包房区，实木板的用餐台，给予客人无比温馨的用餐幸福。

这里仿佛是在与世隔绝的氛围中，厨师制作出色香味齐全的美味，仿佛在瞬间造就了一个世界，将顾客深深地吸引。不必担心满心的欢喜和诱人的美食会悄悄地溜走，人们满足于这充满幸福的时光。

Xijinhui Neo-Chinese Style Restaurant

西津会中餐厅

设计公司：上海欧可建筑装饰设计有限公司
设 计 师：冉昕
参与设计：陈晓东
项目地点：江苏省镇江市西津古渡
项目面积：约 2000 平方米
主要材料：祥云纹雕刻板、荔枝面花岗岩石材、大理石、仿古地板、
　　　　　水磨石地坪、木饰面、软包、壁纸、马来漆等

Xijindu, an old ferry and a fashion attraction that connects Zhenjiang City's history and modern times, and our stories happen here.

In the design for the project, the tourism situation and the features of the Xijindu historic street are considered. The design adopts Neo-Chinese style. It's our focus to combine Neo-Chinese style into the local construction style. The bamboo forests, stream, bricks and white wall come into the inside from the outside. Common materials such as silk, fiber, bamboo, wood create an elegant space by the comparison of heaviness and lightness, brightness and darkness, hardness and softness. The use of Chinese style lattice separates the areas yet they seem to be connected, so the space appears more fluent. The auspicious cloud decoration is matched with the details of dots, lines and planes of the furnitures and the lightings, endowing the space with poetic ambiance

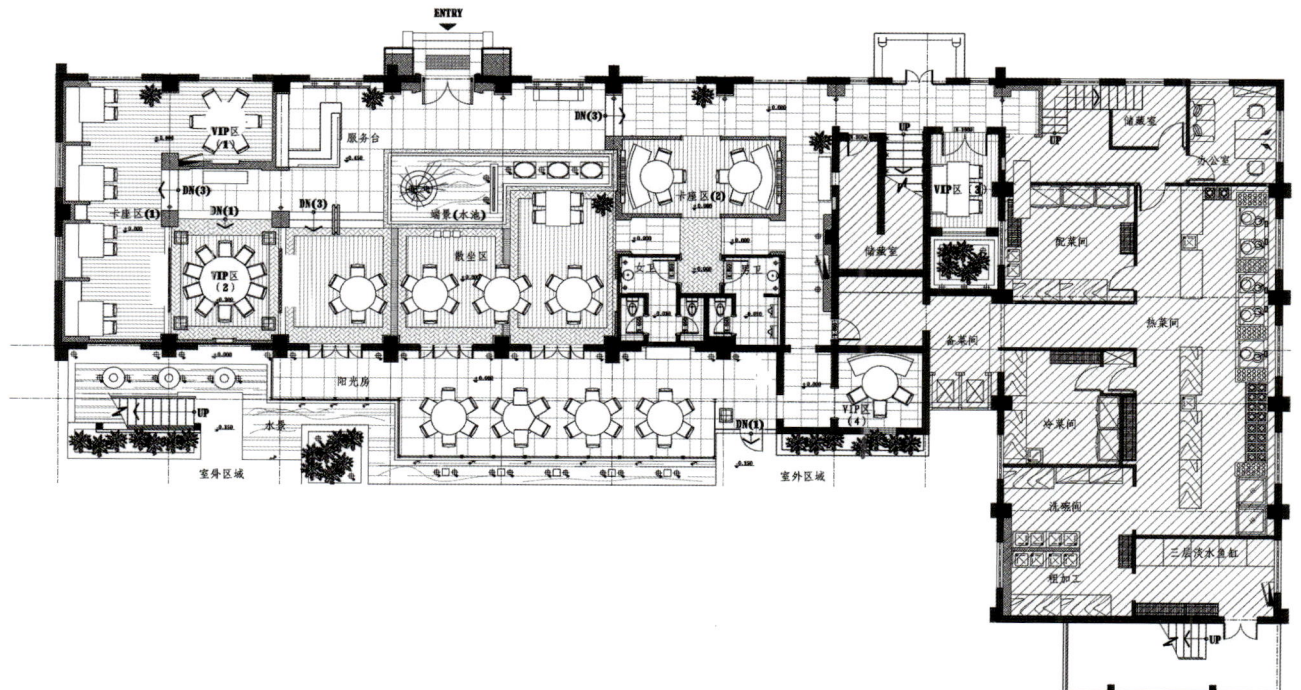

西 津渡，一个古老的渡口，一个将镇江的昨天与今天相串连的时尚景点，我们的故事就在此发生。

整个项目结合了周边旅游业的发展情况和西津渡历史街区的特点来审视和设计，总体设计风格定义为新中式主义。新中国风融入当地建筑风格是我们想要塑造的重点。竹林、流水、青砖、白墙从室外延伸到室内。丝、麻、竹、木等普通材质通过轻重、明暗、刚柔、虚实的对比塑造出高品质感的空间。中式花格的运用将界面隔而不断，使空间秩序更为流畅。祥云图案配以家具、灯具上的细节，以点、线、面造型，赋予空间诗意构图。

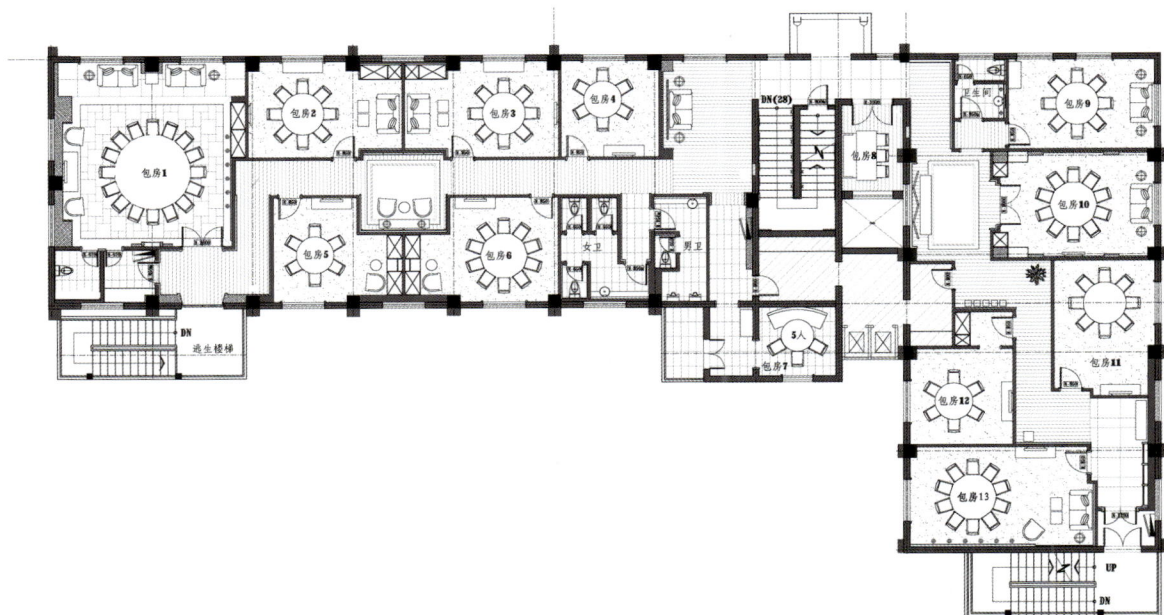

Rehe Restaurant

热河食府

设计公司：北京大石代设计咨询有限公司
设 计 师：霍庆涛
参与设计：赵洪程、汤善胜、丁洁
项目地点：石家庄中山西路
项目面积：2600 平方米

This case is about a Chinese restaurant which mainly serves the royal food beyond the great wall. Using modern means to express Chinese classical connotation is one subject the desinger has been pursuing and practicing. The desinger takes Chengde culture as a background, combining architectures such as Mulan Yards, Chengde Summer Resort, and Stranger Temple to integrate cultural and natural scenery design style to the space. The large Chinese landscape painting, natural landscape settings, blue and white porcelain, Beijing Opera mask, violet arenaceous teapot, chess, colorful butterfly show the 72 scene natural scenery of Chengde Summer Resort.

The interior space is without the slightest redundance. The light environment of "Feel the light, yet can't see the lamps" makes the space impressive, deep, and rich of layers. The use of tawny glass makes the space full of changes and diversified forms of expression.The space gets to extend by "no cost copy" of the decorations. In adornment, the designer used the pictures of a book about Qing emperors and Chengde Summer Resort to decorate the space, which also conveys Chengde culture.

本案是一家经营塞外宫廷菜的中餐厅，用现代手法传达中国古典意蕴是本案设计师一直在探索和实践的一个课题。设计师以承德文化作为大背景，融合木兰围场、避暑山庄、外人庙等建筑，人文与自然风光为一体的设计风格被设计师巧妙地运用到空间中。大幅的中国山水画、自然景观布景、青花瓷器、京剧脸谱、紫砂壶、象棋、彩蝶像一个个建筑精灵，向人们展示着避暑山庄72景的自然风光。简洁的室内空间没有一丝多余的表达。"见光不见灯"的光环境使空间显得深邃而悠远，且富有层次。茶色镜片的使用使空间富有了变化和多样的表情，使空间得到延展的同时也对装饰进行了"无成本复制"。在装饰画的运用上，设计师更是别出心裁地将一本介绍清帝和避暑山庄的书籍拆散进行装裱，在装饰空间的同时更加精确地传达了承德文化。

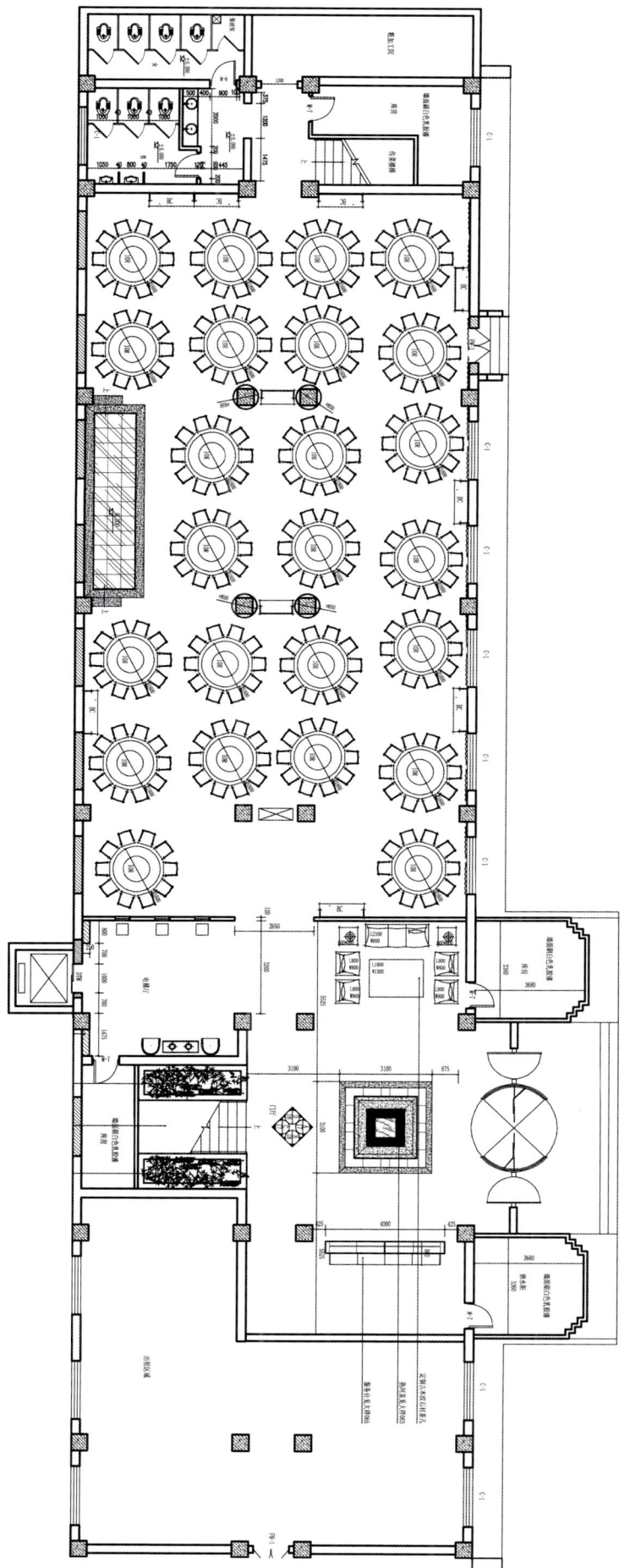

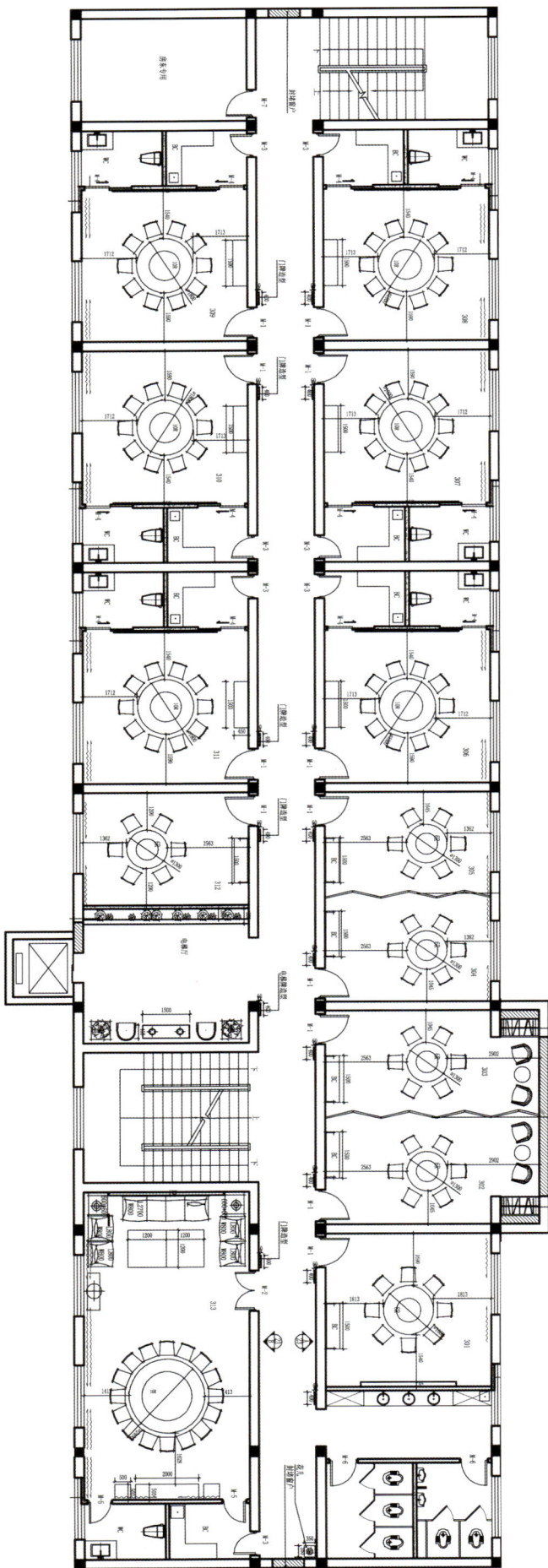

Western
Style

Shanghe Western Restaurant

上荷西餐厅

设计公司：徐庄设计（广州）有限公司
设 计 师：庄仙程
参与设计：周良娇、陈育谦、庄仙任、陈晓佳
项目地点：广州
项目面积：1500 平方米
主要材料：欧亚木纹石、古木纹、灰木纹、墙纸、工艺地毯

In this project, the design not only considers the common steak dining model, but also pays much attention on creating the serving function for activities such as western wedding banquet, news conference, big business party and so on. This is an innovative place of the Western wedding restaurant in Guangzhou.

The T-stage is nearly 40 meters, with professional stage lighting. With the floating-cloud carpet, the space would be an excellent platform for performance.

The changeable lights give customers fresh poetic feeling for dining. As for the food, the fruit Aromatherapy dish is placed in the hall for show. As a result, while enjoying food, the customers can also get the opportunity to appreciate the chef making food.

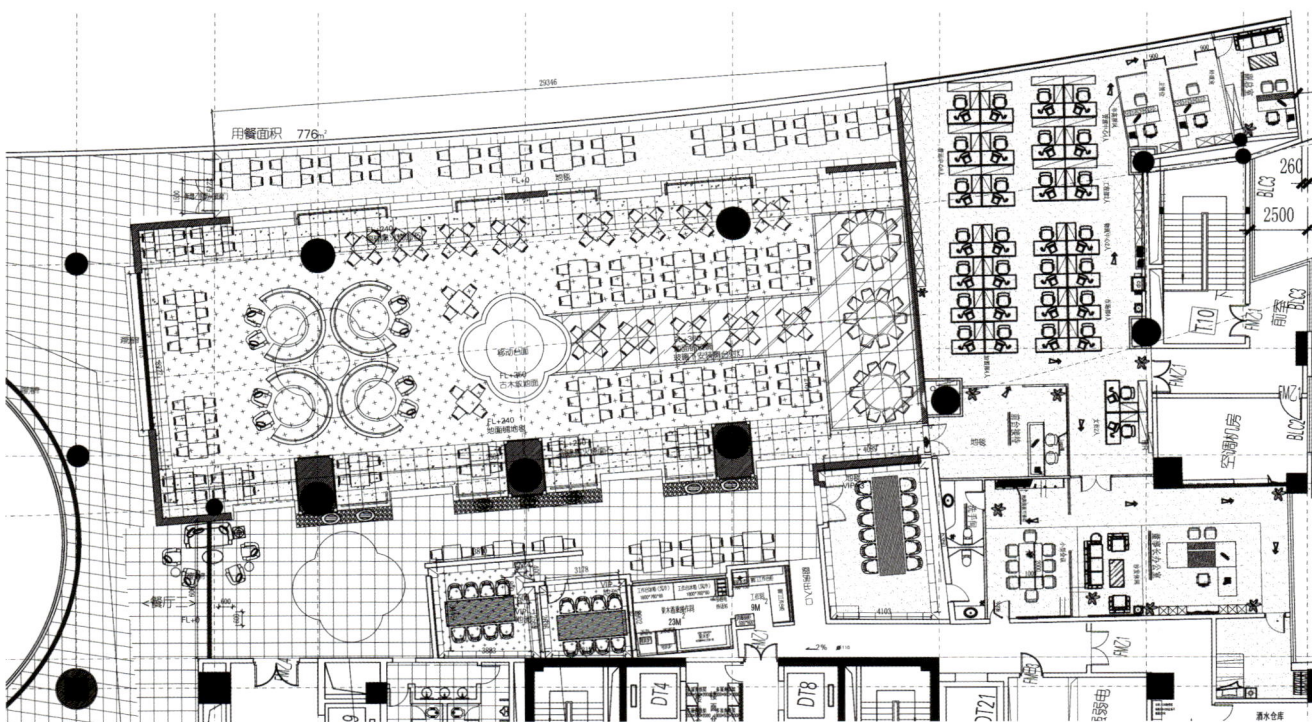

在本案的设计功能上设计师考虑的不单是普通的西餐牛排用餐的模式，更加注重的是西式婚宴、发布会、大型商务聚餐等活动的接待功能，这也是广州西式婚宴设计餐厅的首创之一。

近40米的T台有专业的舞台灯光设计和彩云飘飘的地毯设计，整个空间作为时尚产品发布会的会场将是很好的表演平台。

七彩变幻的灯光，也给客人不一样的用餐诗意。在食物制作功能上更是把果木香熏搬到大厅来展示。客人在享受食物的同时也体验一丝厨师制作的乐趣。

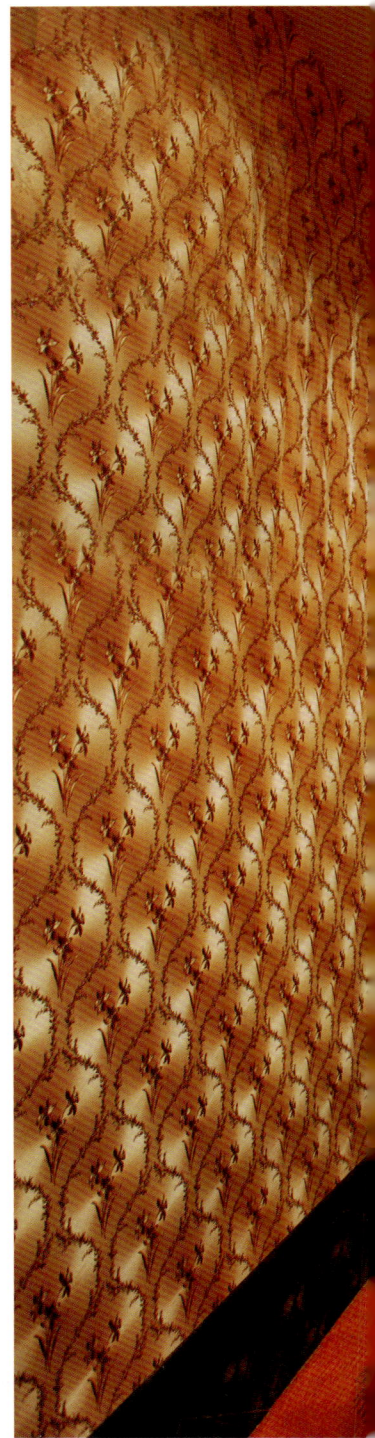

Wuhan Greenery Longyang Shop

武汉绿茵阁龙阳店

设计公司：广州文智设计工程有限公司
设 计 师：韦文生
项目地点：武汉
项目面积：1600 平方米

The whole design of this dining room is like a large artwork, and uses well-crafted technique to create high-brow dining environment. It applies solid wood as the main material, making indoor diffuse fresh and natural breath. With brown as the basic design tone, it reveals a kind of decorous and elegant temperament and involves the dark sofa into one, strengthening the indoor being a sense of whole. In order to avoid excessive dark which brings depressing feeling the stylist applies the white table apperance and sofa cushion, to reconcile indoor atmosphere. It uses hollow carvings to divide the partition areas, which not only increases a visual extension, but also delivers an unspeakable oriental verve and enriches its dining room's taste.

The designs of the top of this space appears so creative, some making people remind of dinosaur spine, some bringing people's thoughts to th ancient manor, and some mixing with different elements, presenting a unique effect...

This kind of restaurant is a pure and quiet land among the rip-roaring city.

整个餐厅的设计就像一个大型的艺术品，用精雕细刻的手法来营造高格调的就餐环境。以实木作为主要材料，让室内弥漫清新的自然气息。以棕色为设计基调，透露出稳重高雅的气质，并与深色系的沙发形成一体，加强了室内的整体感。为了避免过多暗色给人带来沉闷感，设计师将所有的桌面和沙发靠垫都选定为白色，以调和室内气氛。采用镂空雕花隔断来区隔空间，增加视觉穿透感的同时，还传递出不可言喻的东方神韵，丰富了餐厅的内涵。空间顶部的设计显得创意十足，有的让人联想到恐龙的脊骨，有的将人的思绪带到了古老的庄园，有的则将不同的元素混搭在一起，呈现出独一无二的效果……

这样的餐厅，是喧嚣都市中的一片净土。

Wuhan Wuguang Center Lotus Pavilion Restaurant

武汉武广中心荷花亭餐厅

设计公司：广州文智设计工程有限公司
设 计 师：韦文生 韦智生
参与设计：张诗慧 邓全智 杨明艳
项目地点：武汉
项目面积：1600 平方米
主要材料：地砖、斑马饰面板、玻璃
摄　　影：钱翔

The soft light is matched with the zebra pattern veneer, which is the highlight of this restaurant. The simple yet soft color tone and the lines produce strong visual impact; besides, lotus leaves are decorated among them, creating impressive and poetic dining atmosphere. The construction can get daylight well. The window is carefully designed, with exquisite details. The pavilions around are like little paradise for dining, catering for the customers. The designer divides the center of the restaurant into two parts, which provide the customers with more choices. What's more, besides the exquisite dining chairs, some cupboards and plants are placed in the space for decoration, so the environment becomes natural and comfortable, creating a better dining atmosphere.

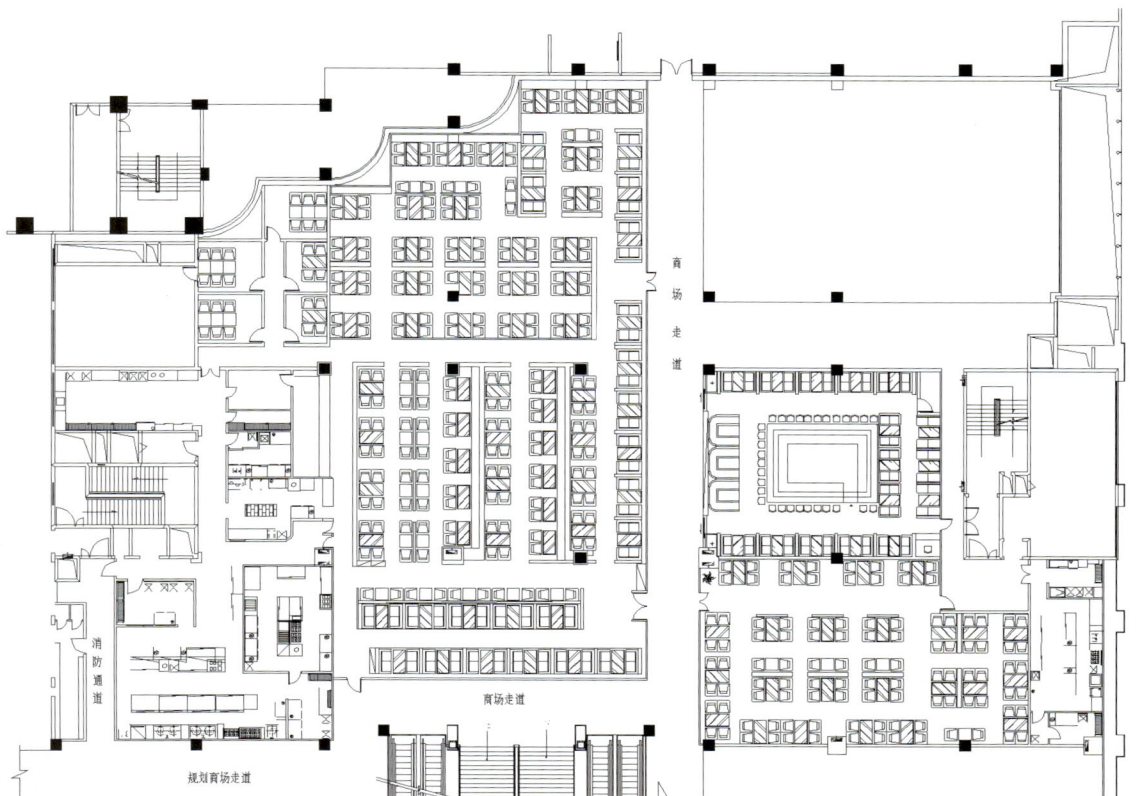

The water bar and kitchen are with simple and bright appearance, which is interesting. The artificial stone veneer is easy to clean, and it forms a comparison with other materials, appearing clear and bright; in the passageway, the designer uses totally-enclosed transparent clear glass, decorated with wooden geometry modeling. The hard glass appears to be flexible floating patterns. In the light, they form rhythm.

One feature of the restaurant is that the serving passage is totally separated from the dining area. The floor is designed to be simple, practical for the dining car.

柔美的灯光配以单一的斑马饰面材料是本餐厅的一大亮点，朴实柔美的色调，线型组合的体量感，使视觉节奏变得更为强烈；荷叶的雅静融于其中，构成了一个既有强烈视觉效应，又有诗意的用餐氛围。

外围运用建筑采光，窗口的造型排列在细节上寻求变化。周边的小亭子，是用餐的小天地，也迎合了客人的喜好。设计师把餐厅中间部分分割成两大区域，人性化的设计也让客人有更多的选择。此外，在比较通透的空间中，除了餐椅布置，也摆放一些柜子、花草等陈设，使室内环境变得自然舒适，也增加用餐情调。

水吧和厨房在外观设计上，造型明快简单，很有趣味性。人造石台面利于打理，同时材质上的反差，更显得清新明快；服务过道中，设计师用了全封闭的透明清玻，嵌上木条几何造型，把生冷的玻璃，软化成漂浮的序列图条，在灯光的映射下，节奏感很强。

区域隔断设置服务走道，与用餐区互不干扰，这是餐厅的一大特点。地面的处理简单、实用，也利于餐车的运作。

Greenery Western Restaurant

绿茵阁西餐厅

设计公司：广州文智设计工程有限公司
设 计 师：韦文生
项目地点：武汉
项目面积：1600 平方米

安全出口➡

In this space with overall round structure, the designer skillfully divides each area, not only embodying a sense of style, but also making full use of interior space. Various lines from the dining room demonstrate themselves by the solid wood and stainless steel material, which not only strengthens the modern and depth of the restaurant, but also brings people a visual impact, building a special dining atmosphere. The decorative use of mirror surface helps to express light shadows and can enrich visual content of space.

For the design of dining room, emotion to foil is very important. The designer here chooses the charming and lovely flowers and sketch paintings which present the diverse and complicated features of human beings, to fill the indoor, which reaches the extraordinary effect.

In a free afternoon, or a quiet night, chat with several friends here, and have delicious meals happily, you will find that life can be so comfortable.

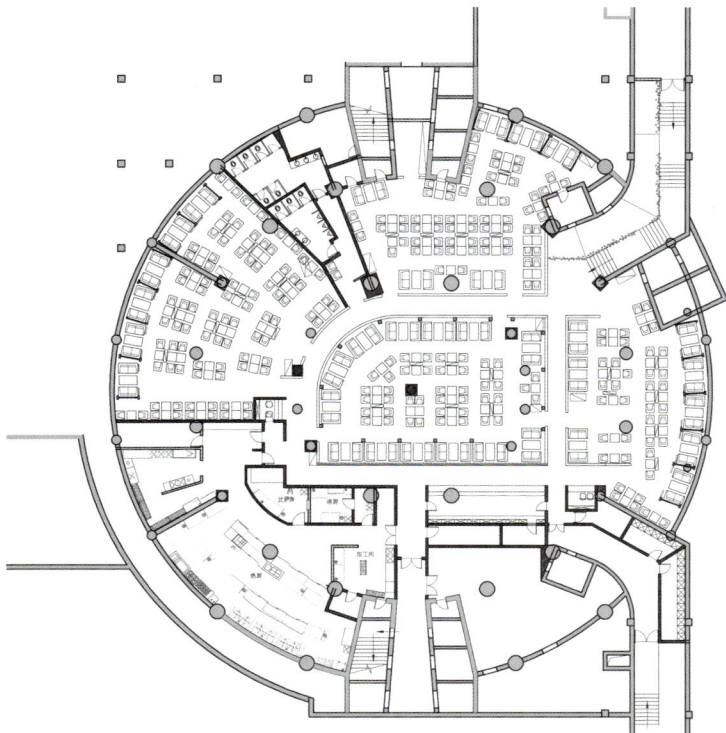

本空间总体结构呈圆形，设计师巧妙划分出各区域，体现设计感的同时，也充分利用了室内空间。餐厅中出现的各种线条借助实木和不锈钢材质表现出来，不仅加强餐厅的现代感和纵深感，还给人的视觉带来冲击，营造出与众不同的就餐氛围。镜面的穿插使用，有助于光影效果的展现，能够丰富空间的视觉内容。

对于餐厅的设计，情调的烘托也很重要，在这里，设计师选用了千娇百媚的鲜花和呈现人间百态的素描画作来填充室内，达到了不俗的效果。

某个悠闲的午后，或是某个宁静的夜晚，约三五好友在此聊天、大快朵颐，你会发现生活原来可以如此惬意。

Kunming Wangfujing Rhine Spring

昆明王府井莱茵春天

设计公司：广州文智设计工程有限公司
设 计 师：韦文生
项目地点：昆明
项目面积：450 平方米

Black and white, which can most highlight a sense of modern fashion being the major tone of the restaurant, combining it with fine materials can quickly manifest style and taste. Refined and unique ornament in indoor everywhere builds a rich emotional dining atmosphere, foiled by the lamplight. The design of hollow solid wood filled with glass, not only makes each functional area separate but link and expands indoor space and brightness, but also diversifies the dining-room expression style and effectively melts the coldness between the black space and the white space which adds some warm to the restaurant. Meanwhile using line flexiblely makes this stylish restaurant have strong artistic breath, and make this static space full of vigor. Dining in this space, people tend to relax their mind during dinner, very suitable for young people's demand nowadays.

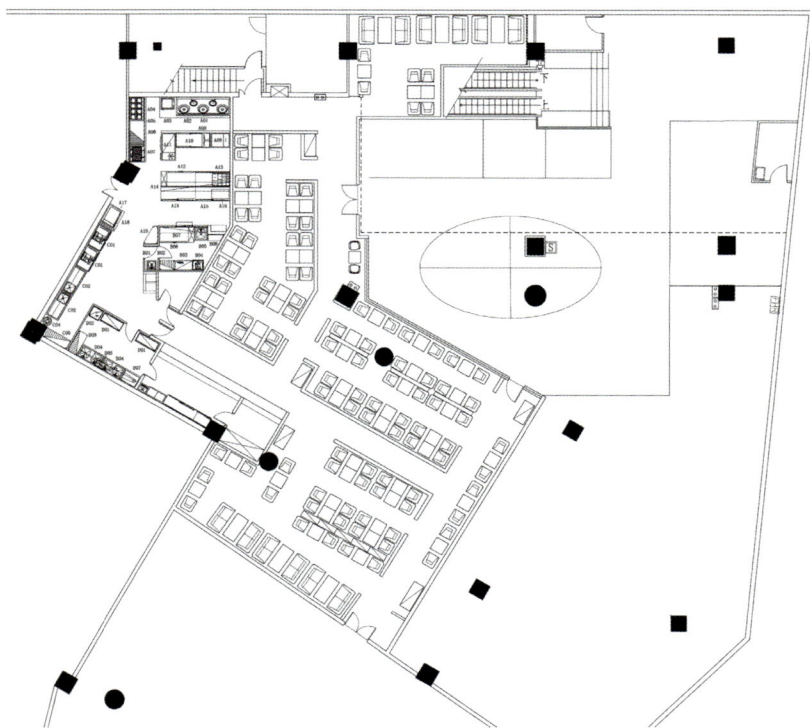

黑与白，最能突显现代时尚之感，将其作为餐厅的主色调，结合优良的材质，格调与品位能够迅速彰显出来。点缀于室内各处的装饰物精致、独特，在灯光的烘托下，营造出富有情调的就餐氛围。实木镂空隔断与玻璃相结合的设计，不但让各功能区隔而不断，增加了室内宽敞感和明亮度，还让餐厅的表现形式更加丰富多样，并且有效地调和了黑白空间引起的冷峻感，为餐厅增添了几分亲和力。同时，线条的灵活运用，让这个现代风格的餐厅拥有了浓厚的艺术气息，也让这个静态的空间变得充满活力，在这样的空间中进餐，自然能够放松心情，十分符合时下年轻人的需求。

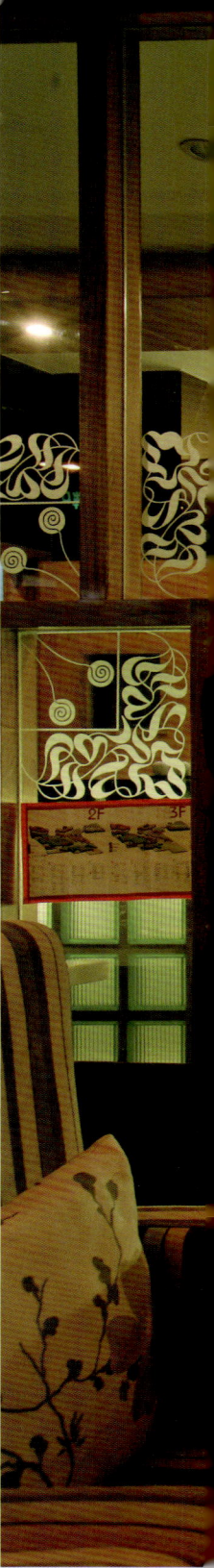

Vogue Coffee

风尚西餐厅

设计公司：上海欧可建筑装饰设计有限公司
设 计 师：陈晓东
项目地点：江苏省镇江市西津渡历史街区
项目面积：约 400 平方米
主要材料：青砖、荔枝面青石板、仿古镜、黑镜、水曲柳木饰面染色、实木地板、
酒红色有机玻璃板、藤面壁纸、墙艺漆

Vogue Coffee is located in the old street of Xi jindu Zhenjiang City. The boss David has been a chef abroad for more than 10 years, and he considers the restaurant his hope, that's why he named it "Vogue Coffee."

As situated in the historical Gudu street, where the constructions retain the simple old natural characteristics. The designer made the best of the advantage, combined Chinese elements and Western elements, as well as western dining culture perfectly eclecticism is embodied fully in this case.

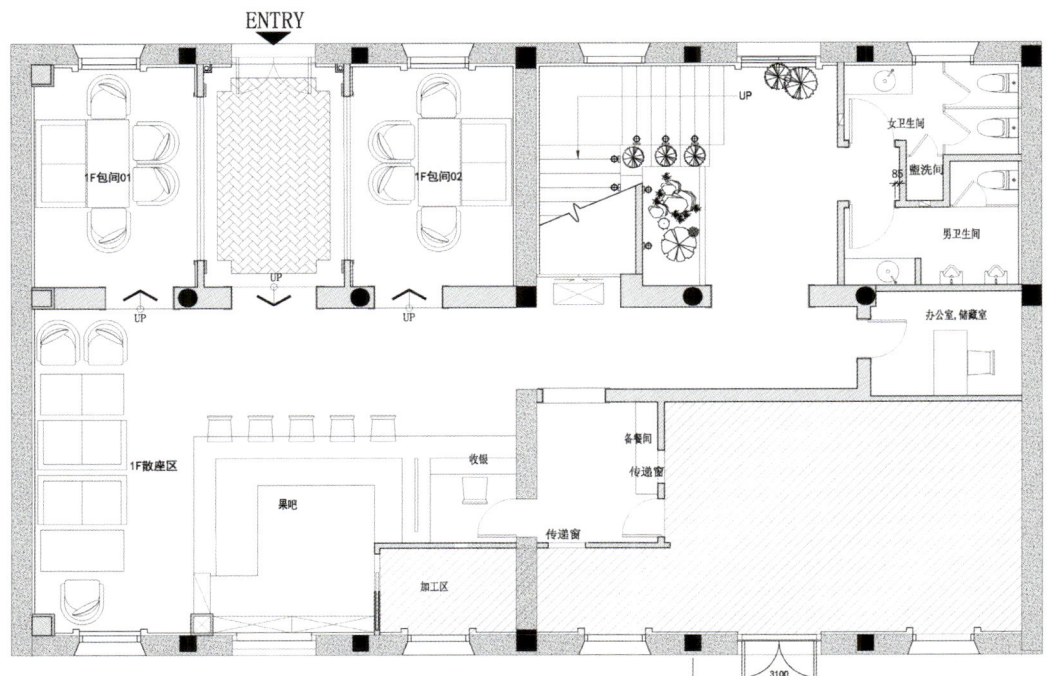

ENTRY

女卫生间

盥洗间

男卫生间

UP

1F包间01

1F包间02

办公室,储藏室

各餐间

传递窗

收银

UP

UP

1F散座区

果吧

传递窗

加工区

3100

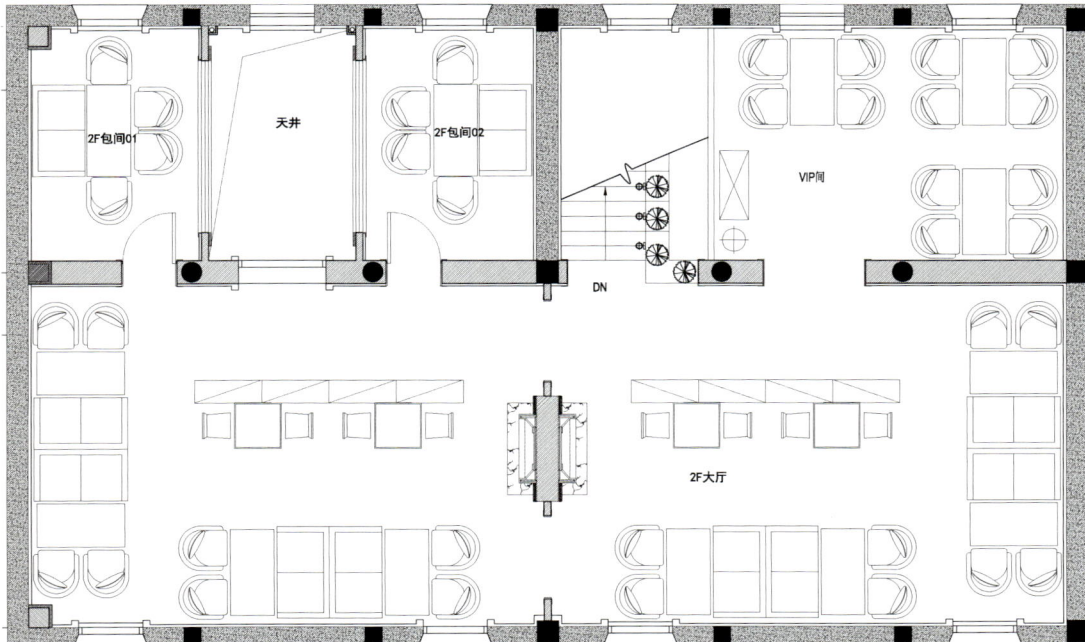

2F包间01 天井 2F包间02 VIP间

DN

2F大厅

风 尚西餐厅，"Vogue Coffee"，坐落于镇江西津渡古街区。餐厅的名称之所以取名"Vogue Coffee"，也正是因为餐厅的老板 David 在国外做了十多年的酒店大厨后把所有的希望赋予自己的这个餐厅。

由于地处历史悠久的古渡街区，建筑的原貌也还保留着原有古朴、天然的气息。设计师充分运用了这一优势，将中式、西式元素、西餐文化巧妙地融合在一起，折中主义风格在本案中已淋漓尽致地得到表现。

Bafenbao Restaurant

八分饱餐厅

设计公司：上海欧可建筑装饰设计有限公司
设 计 师：陈晓东（Nico Chan）
项目地址：江苏省镇江市
项目面积：约 900 平方米
主要材料：青砖、荔枝面青石板、仿古镜、黑镜、水曲柳木饰面染色、实木地板、
　　　　　酒红色有机玻璃板、藤面壁纸、墙艺漆等

This restaurant is a reconstruction project. The structure of the inner space of the old building is very complicated, which caused much difficulty for the designer in the beginning. How to create a totally different restaurant giving people a brand new impression?

The designer divided the space into different parts, separating every dining area with real and unreal partition, making the space more interesting. Besides, the designer used dark light as much as possible to create a romantic and mysterious ambiance.

"A strict owner helps to create outstanding designer." During the designing process, the owner put forward high requirement. On the choice of the furniture as well as lighting decoration, the designer thought over carefully, especially in the customized lightings. One of the highlights of this restaurant is the 9-meter shell droplight, which would be more expensive if assigned to others.

　　本案是老店的拆除改建项目，原建筑内部空间及结构错综复杂，设计伊始也为设计师造成了不少难度。怎样使新店的面貌焕然一新，给人以不同的感觉？首先设计师将内部空间做了不同的划分，对每个用餐区域进行分隔，采用虚实相结合的设计手法，使空间富有变化、耐人寻味。另外，在灯光的控制上，设计师尽量将灯光压暗，制造出一种浪漫、神秘的氛围。

　　"挑剔的业主才能够成就优异的设计师。"设计过程中，业主对设计师提出了很高的要求。在家具、灯饰的选择上，设计师也是煞费苦心，特别是在灯具的定制过程中。本案的设计亮点之一是九米长的贝壳造型吊灯，如外发定做的话其造价非常高昂。

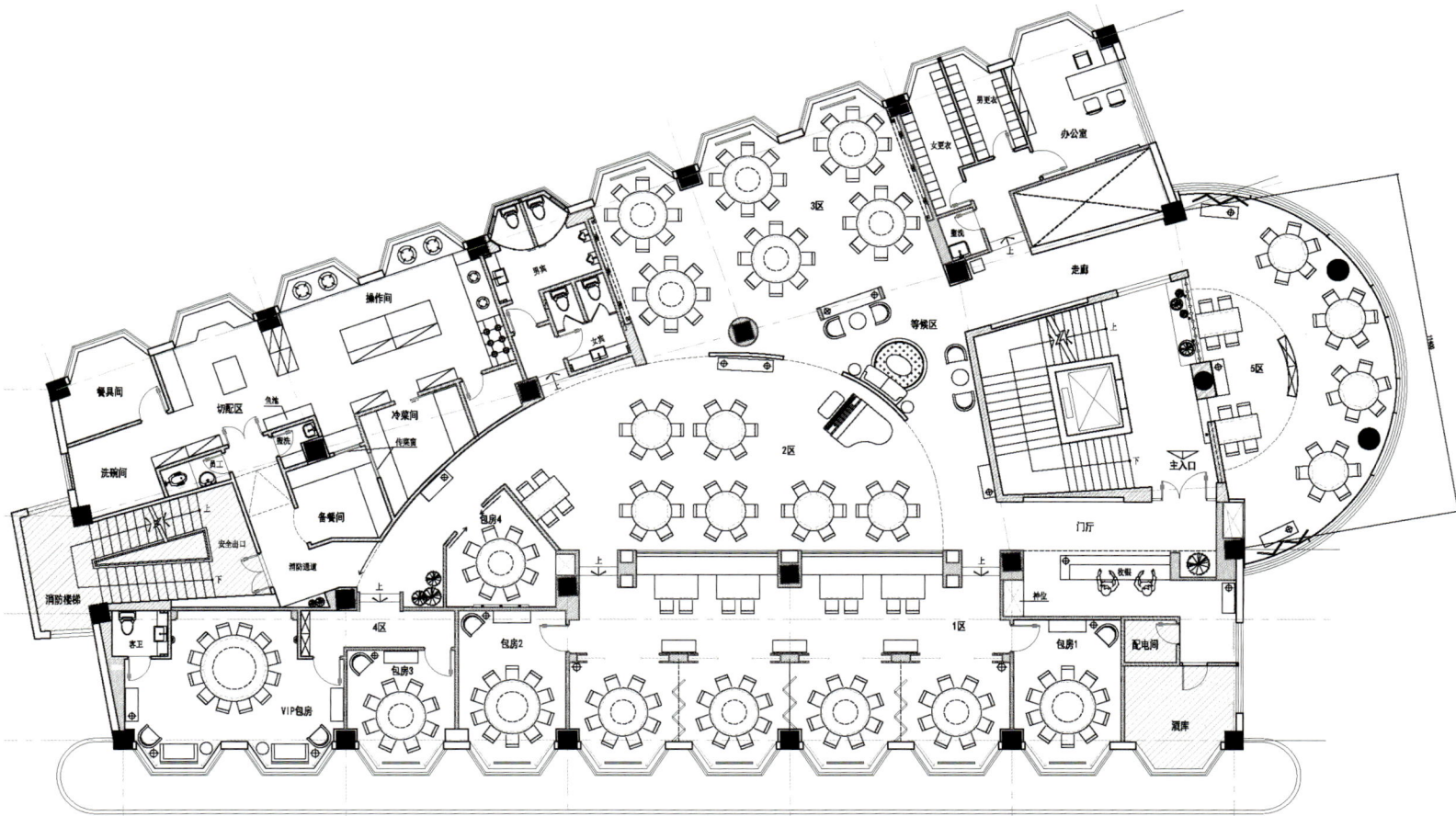

Qinqin Seafood Fine Restaurant

秦秦渔港精品餐厅

设计公司：上海达达室内设计事务所
设 计 师：许清平
项目面积：1400 平方米
主要材料：墙纸、地砖等

This restaurant mainly offers river and sea food. The boss aims to create exquisite catering chain-store, so the design style is modern. The ceiling of the entrance adopts simplified glass concept fish, which echoes to the management direction of the restaurant. Black, white and grey are the domain colors of the restaurant; the furniture is covered with light saturation dark purple fabrics in order to create a fashionable and magnificent restaurant. There are no other mixing colors, in some parts, vibrant colors are adopted to signify the flourishing vitality of the restaurant.

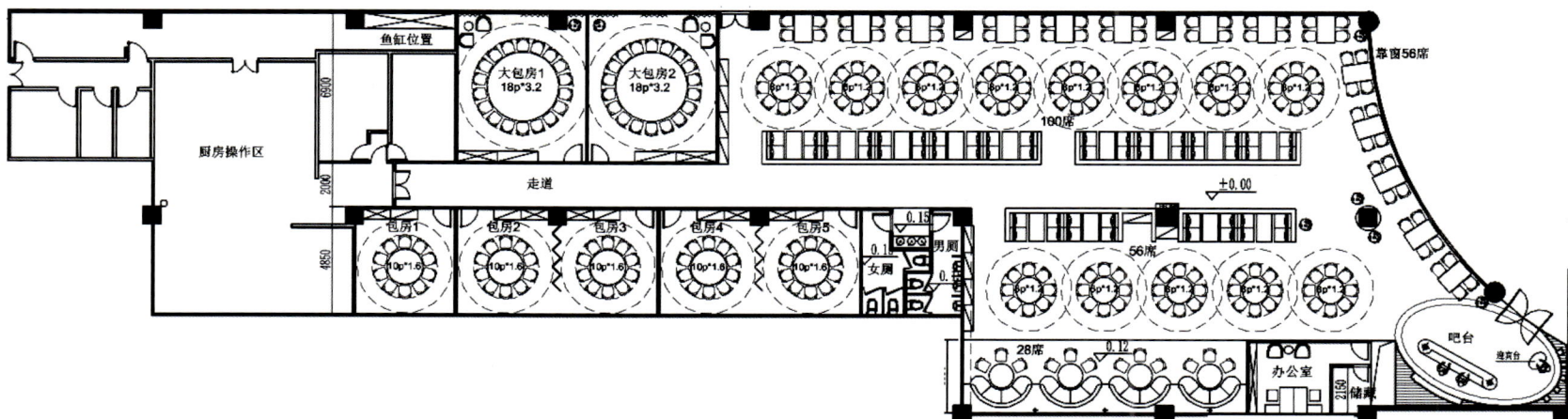

这是一个以经营河鲜、海鲜为主的餐厅，业主尝试打造出精品餐饮连锁店，所以在风格的定位上以现代为主，入口的天花设计上采用简化了的玻璃概念鱼，很好地契合了餐厅的经营方向。整个餐厅的颜色以黑白灰为主色调，家具辅以低彩度的暗紫色面料，力图使整个餐厅时尚而不失大气，没有其他杂乱无章的色调，局部造型上选择了向上的、富有方向性而表达出勃勃生机的概念，寓意餐厅蓬勃发展的生命力。

现代休闲

Modern Recreation

MOJO Icusine Interactive Restaurant

MOJO 交互式潮流概念餐厅

设计公司：木石研室内建筑空间设计有限公司
设 计 师：范赫铄
参与设计：钱欣 林立鑫 陈巨翔
业　　主：申一餐饮有限公司
摄　　影：MARC GERRITSEN

If a beautiful space is deprived of solid materials of which you are familiar to, such as wood, iron, stone and glass, then what is left? Yes! It is only the users of that space that remain. In other words, it is the objects combined with the movement and behavior of the organic entities which create an actual space. If the contents of a space only exist for the users, then the behavior of those users is the most important factor in meeting the needs of that space. While users' behavior in relation to space is an essential element in design strategy. In this unique restaurant design, the client and the designers tried to define and construct a new type of users' behavior in restaurant.

USERS' EXPRESSION

In the construction of a space, the people using the space are the most important part. The expressions from the users also provide most attraction of one space. To design a restaurant which determined to make people happy, the happiness expressions from the customers become the most important value in our design process. Therefore, we convinced the client to minimize the other spatial elements and focusing on the behavior of customers.

TOUCH

MoJo iCuisine is the first realized, fully interactive restaurant in Taiwan China. The moJo iCuisine interactive dining table is a modular table being able to seat two diners. It is equipped with touch sensors and an attractive visual interface. The interface is projected by way of an overhead installation; specifically designed to be viewed from two opposite orientations.

The touch sensors allow diners to interact in several different and interesting ways. For example, diners can touch and toss the circular menu, directly order dishes from the kitchen, change the digital table cloth, view advertisements, play games, fill out opinion forms and check or pay bills. When a user orders a dish, the kitchen will receive it in real time, allowing the chef to make it immediately while concomitantly being charged on the bill. The interactive tables are overlaid with vivid and colorful changing graphics providing for delightful and impressive moments. The interface itself also creates a flourishing and colorful landscape that ensures a memorable dining experience.

INTEGRATION OF VISUAL COMPONENTS

For the implementation of the moJo iCuisine project we designed not only the overall visual appearance which included the entry images, business cards and interactive table but also the establishment of a database system according to the "service design" concept.

The information system linked with the interactive table can be used to record and track consumer preferences, compiling statistics on the operation of the restaurant. The integration of visual elements resulted in the dynamic qualities of space, from static images to an interactive video interface. Biological organisms were also highlighted such as the projection lamps being arranged in the air like a group of organized life.

如果从绚丽的空间里抽掉实体的材料，如你所熟悉的木、铁、石材、玻璃时，剩下的是什么？是的！只剩下使用者。也就是说，空间的存在是空间的实体与使用者两者所构成，而使用者在空间的游移、运动与行为构成了空间的有机要素，如果说空间中的所有内容是为了使用者而存在，那么，使用者的行为就是一个空间中最重要的成分，满足使用者在空间中的需求，能够成为在空间建构中的重要策略要素，因此，在此独特的餐厅规划案中，该案客户与设计者尝试建构一个新的餐厅中的使用者行为。

人跟人之间的互动表情

长期在空间设计的工作中，最难也最少捕捉的就是使用者的表情。从空间的完成开始，使用者的使用行为就是空间完成后最大的依凭，其中使用者的表情，也可以是空间中除了材料外，最大的张力，而一个以打造幸福的互动感为目的的餐厅，如果失去这个价值，其他的要素都将不具意义，因此，设计者与客户合力挑战，取空间要素的极简化，将空间中的主要要素，都还原在关注使用者在空间中的互动行为，期使餐厅所衍生的互动氛围，弥漫在空间中。

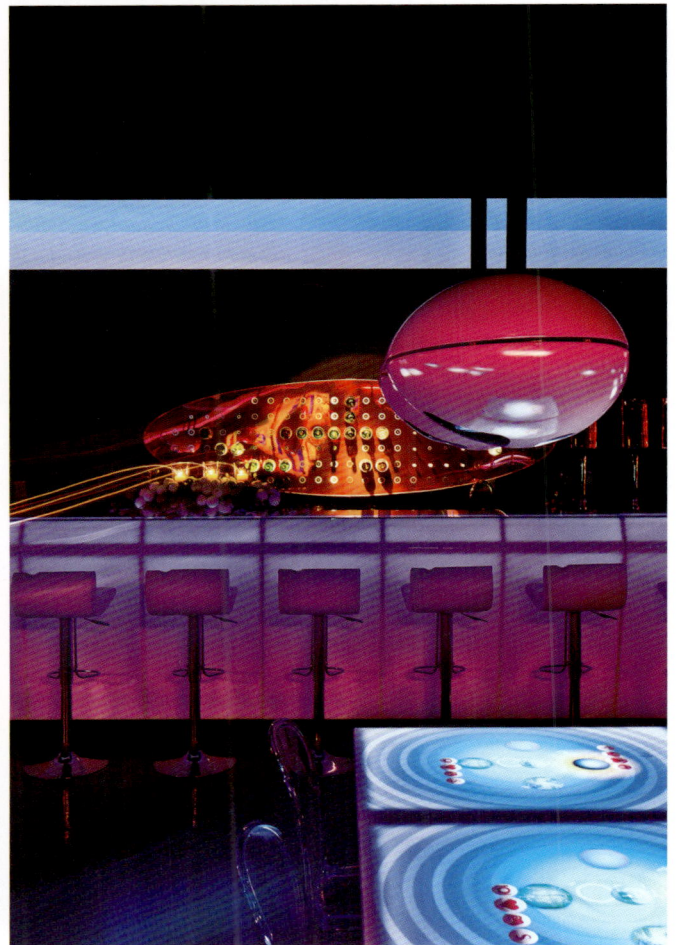

接触

使用者的互动行为，是藉由接触开始，我们关注了使用者在空间的各种接触行为，并且投射在餐厅的各种接口中，从餐厅的主要使用者到服务人员，其中桌面包含餐桌与吧台的触控装置成为主要的聚焦点，moJo iCuisine 互动餐桌是以触控感应装置的模块化餐桌组。其视觉接口由装置上方投射影像成像，并特别设计成可由两个对向观看的接口及互动模式。使用者可触控、抛动环状菜单选择餐点，阅读餐点说明。

除了浏览菜单之外，消费者亦可透过交互式餐桌直接点餐／饮料、等餐时玩小游戏、变更"数字桌面"、观看广告、填写意见表、核对／付账单，且所有动作都在简单的手指点、触间即可完成。餐桌接口本身则充满着色彩丰富的画面，提供消费者一个愉悦的用餐经验。搭配着可口美食和色彩丰富、有变化的互动画面投影，让用餐场景充满了愉悦而惊喜的体验。

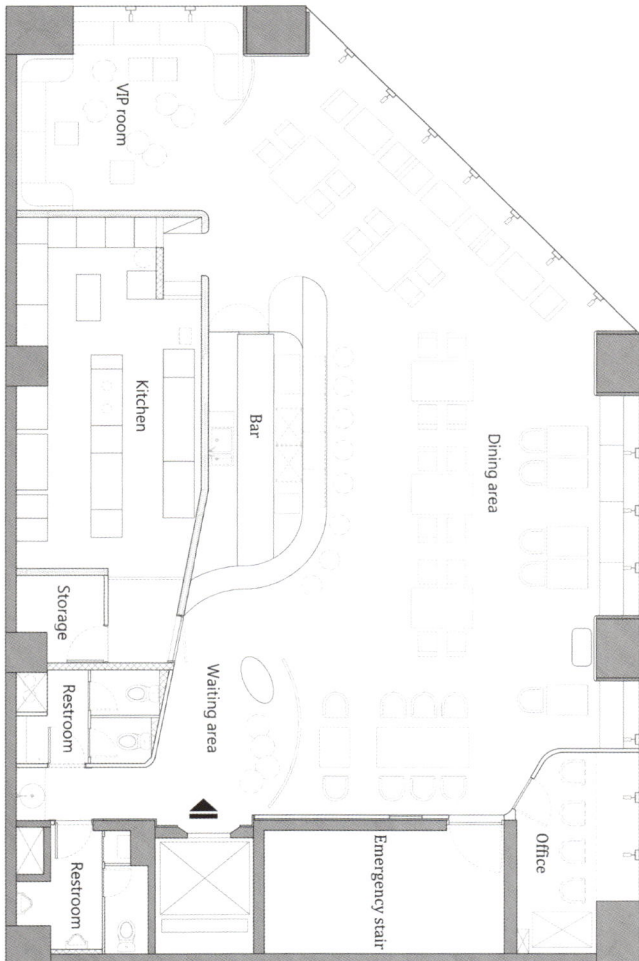

Floor plan labels:

- VIP room
- Kitchen
- Bar
- Dining area
- Storage
- Restroom
- Waiting area
- Restroom
- Emergency stair
- Office

视觉要素统一化

在本项目执行中，我们不仅仅为 moJo iCuisine 设计了整体视觉外观，包含入口影像、名片、互动餐桌的有趣画面，同时也依照"服务设计"的概念，为其建立一套数据库系统。与该数据库系统连接的互动餐桌可记录、追踪消费者的喜好，并汇整出餐厅营运的统计数字。统整后的视觉要素，使得空间的动感特质，由形象识别一直到空间，由平面到空间，由静态影像到互动界面影像，也凸显在像是某种生物有机体的投影吊灯，它们一串串地排列在空气中，像是一套有组织的生命。

Ripple Bed Bar Theme Restaurant

涟漪床吧主题餐厅

设计公司：马迪思·明团设计机构
设 计 师：Manuel Derel、 Eric Wong、 Julian Cornu、 Philippe Colin
项目地点：广州市海珠区滨江东玉菡路 32–34 号珠江新岸酒店首层
项目面积：327 平方米
主要材料：夹板造型环氧树脂漆、墙身乳胶漆

In China, bed bar is a new thing. In a foreign country, bed bar is a kind of social occasion full of fashion, elegance, and health. "Beds" in the regular bed bar are not absolutely the separable beds but the long sofa bed without separation in the middle, where people can lie down comfortably, chat closely, eat and drink, and appreciate the performances.

The purpose of bed bar is to let all people inside it, keep low-key, equal, close to each other as well as relaxed.

Respect each other and enjoy leisure time, and gain the great happiness with people getting along with each other easily. Therefore, it can not only make people relaxed, but also expect people to relax their interpersonal relationship.

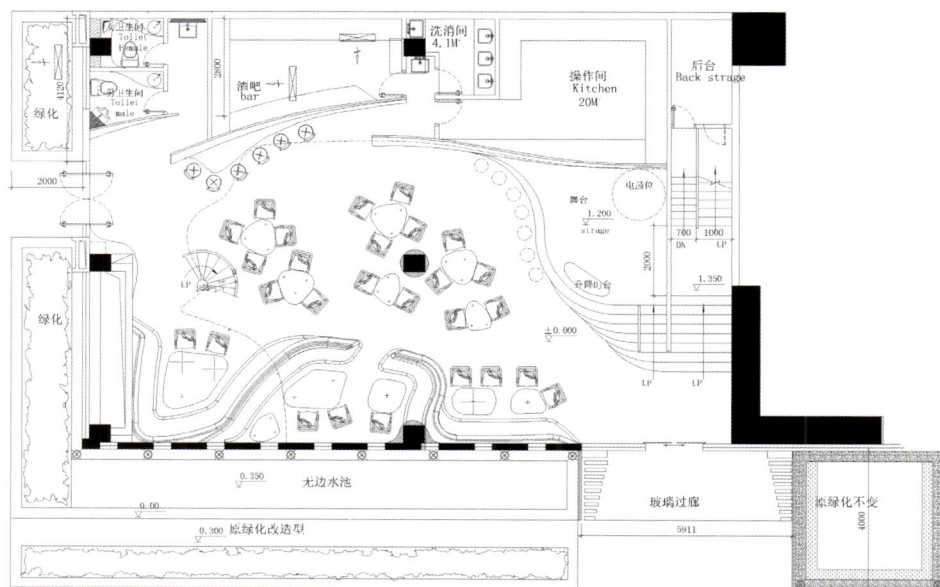

在国内，床吧是个新鲜事物。在国外，床吧是一种时尚、高雅、健康的社交场合。正统床吧里面的"床"绝对不是一张张独立的床，而是中间没有分隔的长长的沙发床，人们可以躺在沙发床上舒服而亲近地聊天、吃喝、欣赏表演。

床吧的精神，就是让所有置身其中的人，都能放下身段，平等、亲近、轻松。

彼此尊重地享受休闲时光、享受人与人之间轻松相处所带来的巨大快乐。所以说，床吧不但能让人们放松身心，而且希望人们放松人际关系。

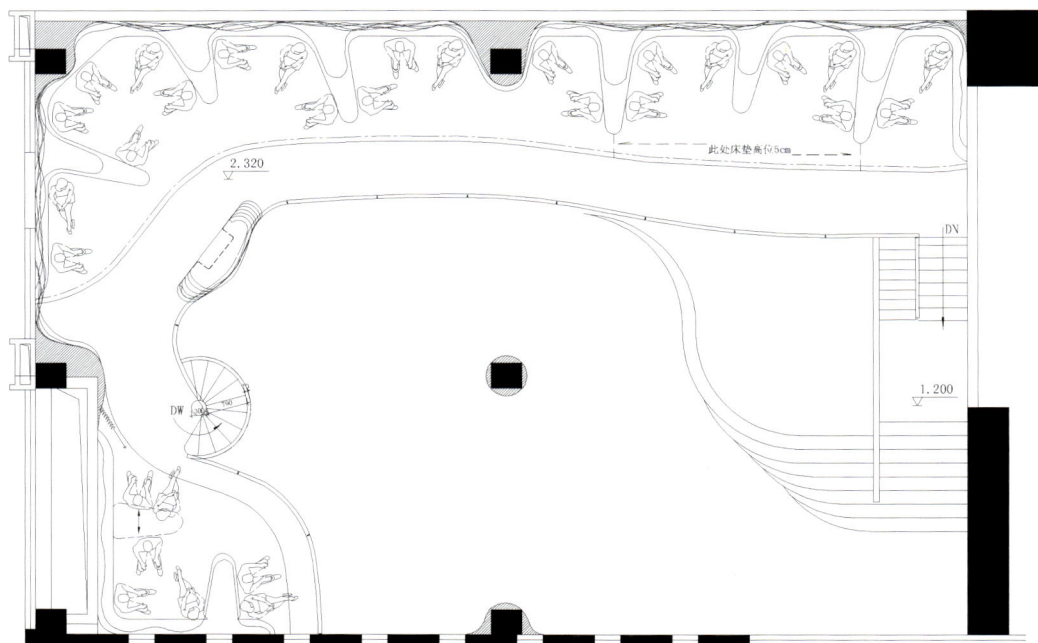

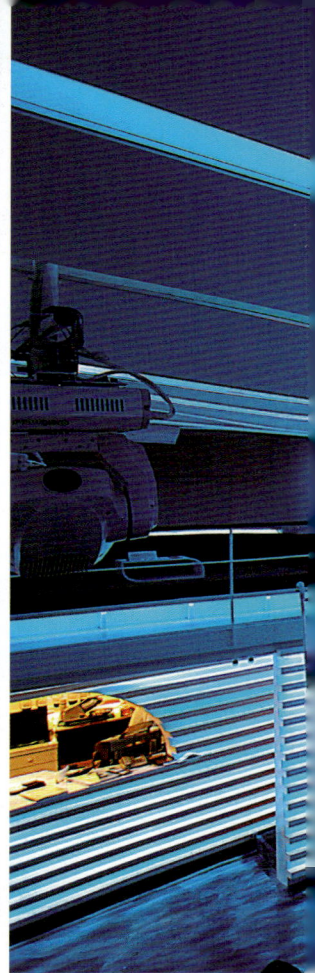

Amazonas Brazil Barbecue Restaurant

亚马迅巴西烤肉餐厅

设计公司：慧驰装饰设计有限公司
设 计 师：欧慧
项目地点：厦门厦禾路益泰大厦店面
项目面积：600 平方米
主要材料：仿古砖、砖纹墙纸、茶镜、不规则板岩
摄　　影：古振宁

This case is a restaurant with natural style, to create the atmosphere of the tropical rainforests, winding rock floor under the high-rising tree or the entrance, cave stairs, the stair armrest in a tangle of branches, as well as the winding beam ceiling showing the whole dining room's move-line, which makes people place themselves into the mysterious tropical rain forests and start their colorful gourmet tour.

本案是一家自然风格餐厅，为营造出热带雨林的气氛，入口的撑天大树下蜿蜒的岩石地板、岩洞楼梯、缠藤树枝的楼梯扶手，同样蜿蜒展现的木梁天花引领整个餐厅的动线，让人仿佛置身在神秘的热带雨林中，开始了丰富多彩的美食之旅。

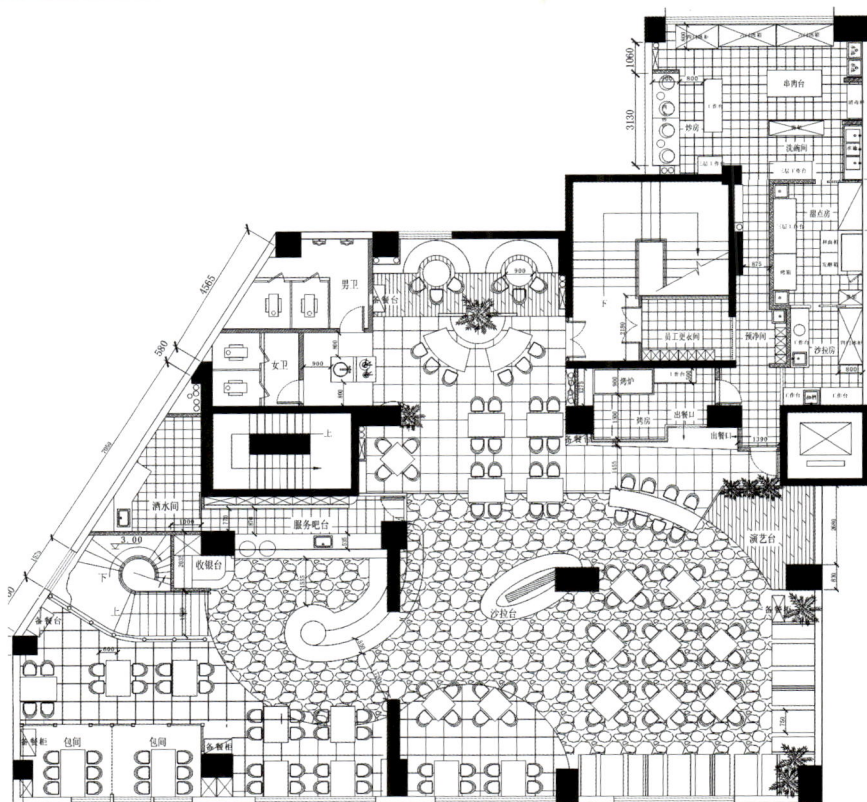

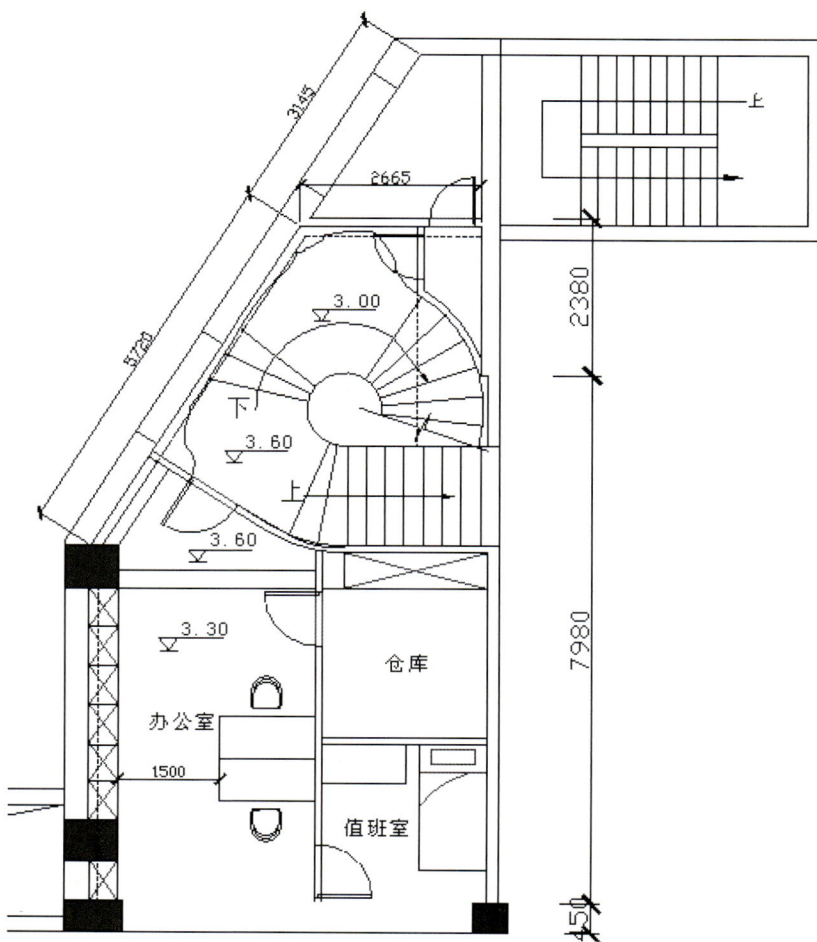

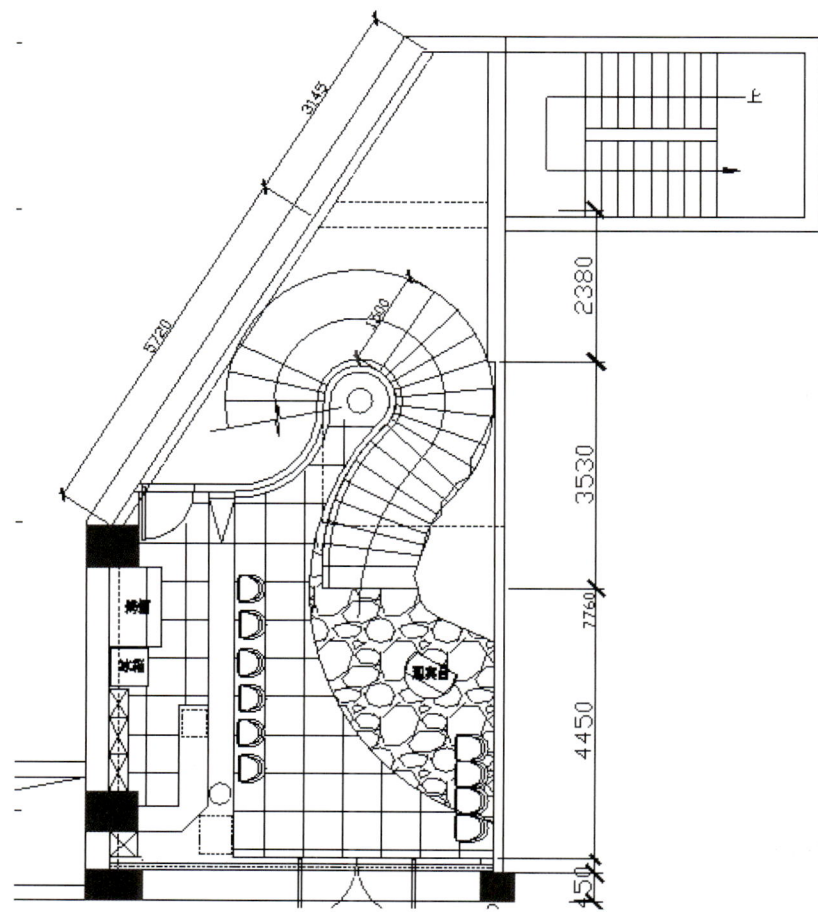

HERE CAFE

哩渡休闲咖啡馆

设计公司：C.DD（尺道）设计师事务所
设 计 师：杨铭斌、李嘉辉、何晓平
项目地点：广东佛山
项目面积：669 平方米
主要材料：木材、红砖、茶灰镜

In the design of this case, the designer tries to complete the architectural space design by the smallest input, based on the traditional design concept. That is to say, in such a good exclusive environment space, not intentionally to decorate the elevation with materials, the tea-grey mirror is installed on the related elevation combined with lamplight, which irradiates the whole space with a kind of gray color. The designer divides the space as if he is cutting the diamond. The symmetry and square interior space is cut into layered functional spaces, breaking the serious feeling of symmetry and square, to create a harmonious comfortable leisure space, all of which makes you being relaxed and charmed inside it with the background music playing……

本案的设计中设计师以传统设计理念为前提，试图以最小的投入完成建筑空间的设计；也就是说，在这样优秀的专属环境空间中，没有刻意将立面用材料去进行装饰，而是结合灯光照射，在相关立面上安装茶灰镜，以一种灰度的色彩映射在整个空间中。设计师对空间的划分就如对着钻石砌割一样，将一个对称方正的室内空间砌割成高低错落的功能空间，打破对称方正的严肃感，营造一个与外部环境融洽的休闲舒适空间，空间的对话使本案空间灵动心扉，在背景音乐下让你如此放松地陶醉于此……

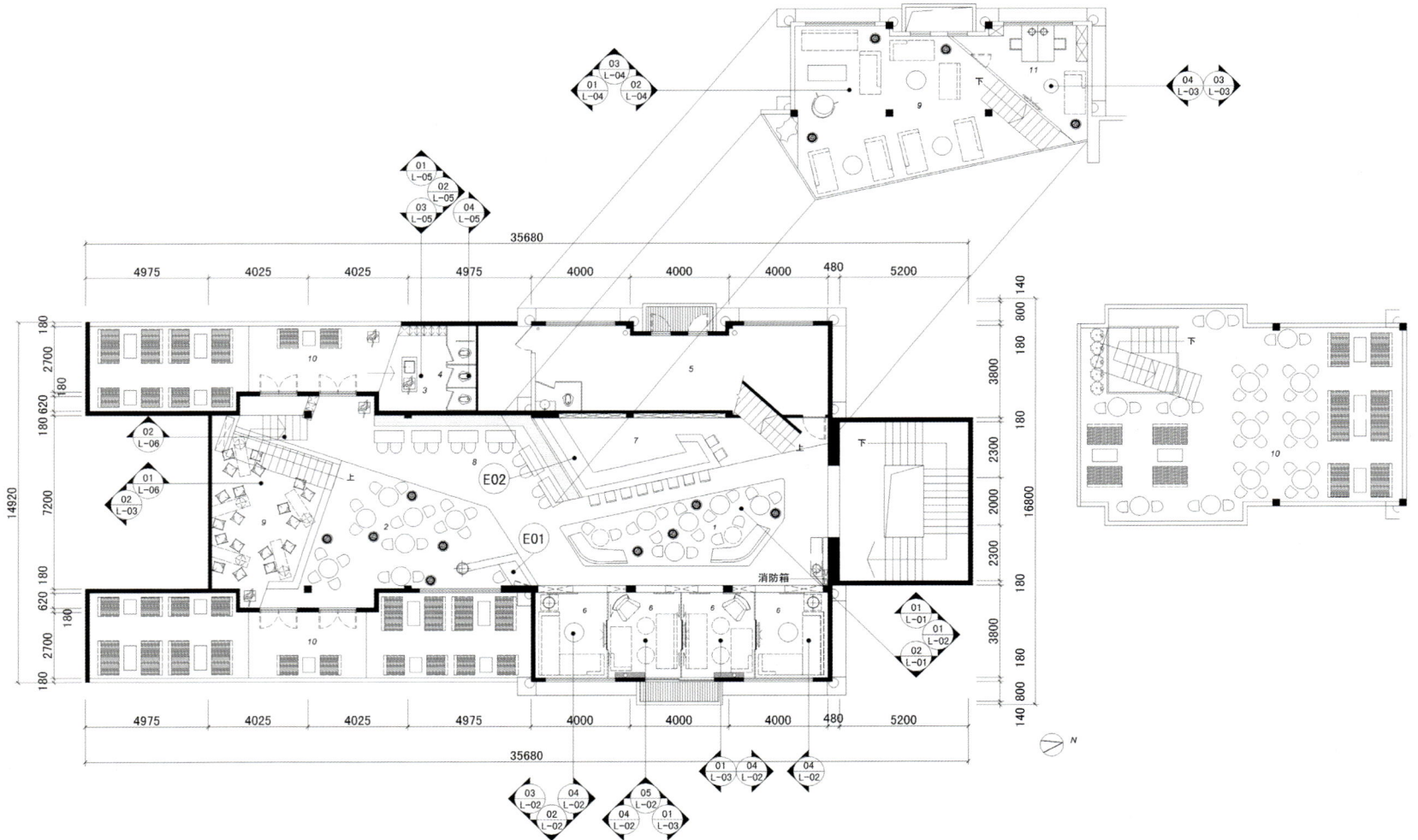

Mingquan Teahouse

铭泉茶馆

设计公司：形上几何策划设计事业
设 计 师：卢迅
项目地点：广东省东莞市南城区八一路
项目面积：850 平方米
主要材料：石膏板（天花）、白色、灰色乳胶漆、水曲柳饰面板（索色显纹）、
　　　　　钢化磨砂玻璃、水泥通花砖

The case is a new orientalistic style, from inside to outside, and attaches much importance to the combination of the tea ceremony spirit and modern civilization tide. The designer manages to put the Chinese elments of deep cultural connotations into interior design. With the popularity of new Chinese style, Chinese culture was excavated out, and passed down to our current life.

本案为新东方主义风格，由内而外，注重中华茶道精神与现代文明潮流的融合。设计师尝试把具有深厚文化底蕴的中国元素融入室内设计中。随着新中式风格的流行，中国文化被重新发掘出来，并延续到我们的现实生活中来。